Lecture Notes in Mathematics

Edited by A. Dold, B. Eckmann and F. Takens

1448

S.K. Jain   S.R. López-Permouth   (Eds.)

Non-Commutative Ring Theory

Proceedings of a Conference held in Athens, Ohio
Sept. 29–30, 1989

Springer-Verlag

Berlin Heidelberg New York London
Paris Tokyo Hong Kong Barcelona

Editors

Surender Kumar Jain
Sergio R. López-Permouth
Department of Mathematics, Ohio University
Athens, Ohio 45701-2979, USA

Mathematics Subject Classification (1980): 16A: 08, 16, 48, 52, 53, 64, 65.

ISBN 3-540-53164-5 Springer-Verlag Berlin Heidelberg New York
ISBN 0-387-53164-5 Springer-Verlag New York Berlin Heidelberg

Printing and binding: Druckhaus Beltz, Hemsbach/Bergstr.
2146/3140-543210 – Printed on acid-free paper

PREFACE

On September 29 and 30 of 1989, Ohio University hosted the Midwest Non-Commutative Ring Theory Conference. With the support of the Office of the Provost for Graduate Programs and Research, the College of Arts and Sciences and the Departments of Mathematics and Computer Science, we were able to bring together speakers from throughout the United States, Canada and the United Kingdom. This volume consists primarily of papers presented by our invited speakers. Some authors who were originally invited but could not attend have also contributed to these proceedings.

While non-commutative ring theory is a vast subject, we believe that there are strong relations among its branches. The Ohio University conference was organized keeping in mind our intention to emphasize and revitalize the interaction between researchers throughout the ring theoretic spectrum. We were left with a good feeling about the unity of our discipline after the conference; we hope that these proceedings convey the same feeling to the reader.

We thank all participants, contributors and referees for their prompt responses which allowed us to stay within a tight schedule bringing this project to its completion. Many thanks are due to Dr. T. Lloyd Chesnut, Associate Provost for Graduate Programs and Research; Dr. F. Donald Eckelmann, Dean of the College of Arts and Sciences; and Dr. Shi-liang Wen and Dr. Klaus Eldridge, Chairmen of the Departments of Mathematics and Computer Science, for their kind support of this conference. Our graduate student, Mr. Abdullah Al-Huzali, helped in many more ways that we can attempt to list here. We would like to take this opportunity to express our appreciation to him. Finally, Ms. Cindy White is to be commended for the excellent job of typing.

The Ohio University conference was dedicated to Professor Goro Azumaya on the occasion of his 70th Birthday. We are happy to dedicate these proceedings to him.

S.K. Jain and Sergio R. López-Permouth
Athens, Ohio, June 1990

CONTENTS

LOCALLY SPLIT SUBMODULES AND MODULES
WITH PERFECT ENDOMORPHISM RINGS

Goro Azumaya

Let M be a left module over a ring R. A submodule N of M is called *locally split* in M if for each $x \in N$ there exists a homomorphism $h : M \to N$ such that $h(x) = x$. It is shown that if N is locally split in M then for any finite number of $x_i \in N$ we can find a homomorphism $h : M \to N$ such that $h(x_i) = x_i$ for all i. The concept of locally split submodules was introduced by Ramamurthi and Rangaswamy [5] by the name of strongly pure submodule and studied in connection with strongly absolutely pure modules. It is known that every locally split submodule of M is a pure submodule of M. The converse is not always true, but this is true if M is projective, as is well known (Cf. Azumaya [2] for these). On the other hand, every direct summand of M is clearly locally split in M. The converse is not true either, and indeed if every locally split submodule of M is a direct summand of M then M must be a direct sum of indecomposable submodules. Moreover, it is a theorem of Bass that R is a left perfect ring, *i.e.*, every left R-module has a projective cover if and only if for any projective left R-module M every locally split (= *pure*) submodule of M is a direct summand of M, or equivalently, the same condition holds for the countably generated free left R-module $M = R^{(N)}$. It is to be pointed out that if the submodule N of M is a direct sum of submodules N_i and if every finite partial sum of $N = \oplus N_i$ is a direct summand of M then N is locally split in M. Thus, it follows in particular that for any family $\{M_i\}$ of left R-modules the direct sum $\oplus M_i$ is always a locally split submodule of the direct product $\prod M_i$.

We now consider the case where M is a finitely generated left R-module and besides M is an R-S-bimodule with another ring S. Let U and V be a left R- and a left S-module, respectively. We say that they form an M-pair if $V = Hom_R(M, U)$ and $U = M \otimes_S V$, where the multiplication $\otimes$ which defines the tensor product $M \otimes_S V$ is given by $x \otimes f = f(x)$ for $x \in M$ and $f \in V$. We shall show that if U and V form an M-pair then there is a one-to-one correspondence between locally split submodules U_0 of U and locally split submodules V_0 of V by associating U_0 with $Hom_R(M, U_0)$ and V_0 with $M \otimes_S V_0$, so that the corresponding U_0 and V_0 form an M-pair. A typical sample of this situation is given by the obvious case where S is the endomorphism ring of the (finitely generated) left R-module M and $U = M$, whence $V = S$, or more generally, $U = M^{(I)}$ and $V = S^{(I)}$ for an arbitrary set I. Anderson and Fuller succeeded in obtaining in [1], Theorem 29.5 a characterization of the left perfectness of S in terms of the decomposability of $M^{(I)}$ that complements direct summands, by transferring this condition to $S^{(I)}$. By employing the same idea and by applying the one-to-one correspondence between locally split submodules established above, we can give another characterization of the perfectness of S by proving the equivalence of the following conditions: (1) S is left perfect, (2) every locally split submodule of the countable direct sum $M^{(N)}$ of M is a direct summand of $M^{(N)}$, and (3) for every set I, every locally split submodule of $M^{(I)}$ is a direct summand of $M^{(I)}$.

1. Let R and S be rings, and let M be a fixed R-S-bimodule. Let U be a left R-module. If we put $V = Hom_R(M, U)$, then V becomes a left S-module by setting $(sf)(x) = f(xs)$ for $s \in S$, $f \in V$ and $x \in M$. Thus if we define $x \otimes f = f(x)(\in U)$ for $x \in M$ and $f \in V$, then we have an S-bilinear multiplication $\otimes$. We call U an *M-coreflexive* module if $M \otimes_S V = U$ with respect to this multiplication.

PROPOSITION 1. Let U be an M-coreflexive left R-module, and let $V = Hom_R(M, U)$. Let D and E be the endomorphism rings of U and V, respectively. Then by associating each $\varphi \in D$ with $Hom(M, \varphi) \in E$ we have a ring-isomorphism $D \to E$, and the inverse of this isomorphism is given by associating each $\psi \in E$ with $M \otimes \psi \in D$.

Proof. That the mapping $\varphi \mapsto Hom(M, \varphi)$ and $\psi \mapsto M \otimes \psi$ give, respectively, ring-homomorphisms $D \to E$ and $E \to D$ follows from the fact that $Hom(M, \varphi)$ and $M \otimes$ are covariant additive functors R-Mod $\to$ S-Mod and S-Mod $\to$ R-Mod, respectively. Let now φ be an endomorphism of U, and let $\psi = Hom(M, \varphi)$ be the corresponding endomorphism of V. Then $\psi(f) = \varphi \circ f$ for all $f \in V$. Thus for any element $x \otimes f = f(x)$ of $U = M \otimes_S V (x \in M, f \in V)$, we have $\varphi(x \otimes f) = \varphi(f(x)) = (\psi(f))(x) = x \otimes \psi(f)$. This shows that $\varphi = M \otimes \psi$ since every element of $U = M \otimes_S V$ is expressed as a finite sum of elements of the form $x \otimes f$ with $x \in M$ and $f \in V$. Let next ψ be an endomorphism of V, and let $\psi = M \otimes \varphi$ be the corresponding endomorphism of U. Then for any $f \in V$ and $x \in M$ we have $(\psi(f))(x) = x \otimes \psi(f) = (M \otimes \psi)(x \otimes f) = \varphi(x \otimes f) = \varphi(f(x))$. This implies that $\psi(f) = \varphi \circ f$ for all $f \in$ and, therefore, $\psi = Hom(M, \varphi)$. Thus the proof is completed.

Theorem 2. Let M be an R-S-bimodule such that as a left R-module M is finitely generated. Let U be an M-coreflexive left R-module, and let $V = Hom_R(M, U)$. Then there is a one-to-one correspondence between locally split submodule U_0 of U and locally split submodules V_0 of (the left S-module) V by means of the following relation:

$$U_0 = M \otimes_S V_0 , \qquad V_0 = Hom_R(M, U_0) .$$

Proof.

(1) Let U_0 be a locally split submodule of U. Then $Hom_R(M, U_0)$ can be regarded as a submodule of $V = Hom_R(M, U)$ in the natural manner. Let f be in $Hom_R(M, U_0)$. Let $x_1, x_2, \cdots, x_m$ be finite generators of the left R-module M, i.e., $M = Rx_1 + Rx_2 + \cdots + Rx_m$. Since $f(x_i)$ is in U_0 for each i, there exists a homomorphism $\varphi : U \to U_0$ such that $\varphi(f(x_i)) = f(x_i)$ for $i = 1, 2, \cdots, m$. Since $x_i's$ are generators of M, this implies that $\varphi(f(x)) = f(x)$ for all $x \in M$, i.e., $\varphi \circ f = f$. Let now $\psi = Hom(M, \varphi) : V = Hom_R(M, U) \to Hom_R(M, U_0)$. Then $\psi(f) = \varphi \circ f$ and so $\psi(f) = f$, which shows that $Hom_R(M, U_0)$ is locally split in V.

(2) Let V_0 be a locally split submodule of V. Then V_0 is a pure submodule of V and therefore $M \otimes_S V_0$ can be regarded as a submodule of $U = M \otimes_S V$ in the natural manner. Let u be any element of $M \otimes_S V_0$. Then u can be expressed as $u = \Sigma y_j \otimes g_j$ with

a finite number of $y_j \in M$ and $g_j \in V_0$. There exists then a homomorphism $\psi : V \to V_0$ such that $\psi(g_j) = g_j$ for all j. Let now $\varphi = M \otimes \psi : U = M \otimes_S V \to M \otimes_S V_0$. Then we have $\varphi(u) = \Sigma y_j \otimes \psi(g_j) = \Sigma y_j \otimes g_j = u$. This shows that $M \otimes_S V_0$ is locally split in U.

(3) Let U_0 be a locally split submodule of U. Let $V_0 = Hom_R(M, U_0)$. Then, according to (1), V_0 is a locally split and hence a pure submodule of V. Therefore, we can regard $M \otimes_S V_0$ as a submodule of $U = M \otimes_S V$. Let $x \in M$ and $g \in V_0$. Then $x \otimes g = g(x)$ is in U_0 (because $V_0 = Hom_R(M, U_0)$). Since, however, every element of $M \otimes_S V_0$ is a finite sum of elements of the form $x \otimes g$ $(x \in M, g \in V_0)$, it follows that $M \otimes_S V_0 \subset U_0$.

Let u be any element of U_0. Since U_0 is locally split in U, there exists a homomorphism $\varphi : U \to U_0$ such that $\varphi(u) = u$. let $\psi = Hom(M, \varphi) : V \to V_0$. Then we have $\varphi = M \otimes \psi : U = M \otimes_S V \to M \otimes_S V_0$ by the preceding proposition. This implies that $u = \varphi(u)$ is in $M \otimes_S V_0$. Thus we know that $U_0 \subset M \otimes_S V_0$ and therefore $U_0 = M \otimes_S V_0$.

(4) Let V_0 be a locally split submodule of V. Let $U_0 = M \otimes_S V_0$. Since V_0 is pure in V, we can regard U_0 as a submodule of $U = M \otimes_S V$ (and indeed U_0 is locally split in U as seen in (2)). Let g be in V_0. Then for each $x \in M$ we have $g(x) = x \otimes g \in M \otimes_S V_0 = U_0$ and, therefore, g is in $Hom_R(M, U_0)$. Thus we know that $V_0 \subset Hom_R(M, U_0)$.

Let f be an element of $Hom_R(M, U_0)$. Then $f(x)$ is in $U_0 = M \otimes_S V_0$ for all $x \in M$. Let $\{x_i | 1 \leq i \leq m\}$ be finite generators of the left R-module M and $\{y_j | j \in J\}$ (not necessarily finite) generators of the right S-module M. Then for each i we have $f(x_i) = \Sigma_j y_j \otimes g_{ij}$ with $g_{ij} \in V_0$ such that $g_{ij} = 0$ for almost all $j \in J$. Thus there exists a homomorphism $\psi : V \to V_0$ such that $\psi(g_{ij}) = g_{ij}$ for all i, j. Let $\varphi = M \otimes \psi : U = \otimes_S V \to U_0 = M \otimes_S V_0$. Then we have $\psi = Hom(M, \varphi) : Hom_R(M, U) \to Hom_R(M, U_0)$ by Proposition 1. It follows then that $\psi(f) = \varphi \circ f$. On the other hand, we have that $\varphi(f(x_i)) = (M \otimes \psi)\Sigma y_j \otimes g_{ij} = \Sigma y_j \otimes g_{ij} = f(x_i)$ for $i = 1, 2, \ldots, m$. Since $x_i's$ generate the left R-module M, this implies that $\varphi(f(x)) = f(x)$ for all $x \in M$, or equivalently, $\varphi \circ f = f$. Therefore, we know that $f = \psi(f)$ is in V_0. Since this is true for all $f \in Hom_R(M, U_0)$, it follows that $Hom_R(M, U_0) \subset V_0$ and so $Hom_R(M, U_0) = V_0$. Thus the proof of our theorem is completed.

COROLLARY 3. If an R-S-bimodule M is finitely generated as a left R-module and U an M-coreflexive left R-module. Then every locally split submodule of U is M-coreflexive too.

2. Let M be an R-S-bimodule. Suppose there is given a family $\{U_i | i \in I\}$ of M-coreflexive left R-modules, and let $V_i = Hom_R(M, U_i)$ for each $i \in I$. Then each V_i is a left S-module and $M \otimes_S V_i = U_i$ with respect to the multiplication $\otimes$ defined by $x \otimes f_i = f_i(x)$ for $x \in M$ and $f_i \in V_i$. Consider now direct sums $\oplus V_i$ and $\oplus(M \otimes_S V_i) = \oplus U_i$. As is well known, we have $M \otimes_S (\oplus V_i) = \oplus U_i$ with respect to the multiplication $\otimes defined$ by $x \otimes (f_i) = (x \otimes f_i) = (f_i(x))$ for $x \in M$ and $(f_i) \in V_i$. On the other hand, suppose that either I is finite or M is finitely generated as an R-module. Then there is a canonical S-isomorphism $Hom_R(M, \oplus U_i) \to \oplus V_i$, which

is given by associating $f \in Hom_R(M, \oplus U_i)$ with $(f_i) \in \oplus V_i$ by means of the equality $f(x) = (f_i(x))$ for $x \in M$. Therefore, we have $M \otimes_S Hom_R(M, \oplus U_i) = M \otimes_S (\oplus V_i)$ by defining $x \otimes f = x \otimes (f_i)$. But since $M \otimes_S (\oplus V_i) = \oplus U_i$ and $x \otimes (f_i) = (f_i(x))$, we have $M \otimes_S Hom_R(M, \oplus U_i) = \oplus U_i$ and $x \otimes f = (f_i(x)) = f(x)$. This shows that U_i is M-coreflexive. Thus we have

PROPOSITION 4. Let M be an R-S-bimodule, and let $\{U_i | i \in I\}$ be a family of M-coreflexive left R-module. Then their direct sum $\oplus U_i$ is M-coreflexive too provided if either I is finite or M is finitely generated as a left R-module.

We now assume that M is a finitely generated left R-module and S the endomorphism ring of M. Then M becomes an R-S-bimodule, and moreover M itself is an M-coreflexive left R-module, because in this case $S = Hom_R(M, M)$ and $M \otimes_S S = M$ with respect to the multiplication $x \otimes s = xs$ for $x \in M$ and $s \in S$. Therefore, it follows from Proposition 4 that the direct sum $M^{(I)}$ is M-coreflexive for every set I. Now $Hom_R(M, M^{(I)})$ is isomorphic to $S^{(I)}$ as left S-modules by associating $f \in Hom_R(M, M^{(I)})$ with $(s_i) \in S^{(I)}$ such that $f(x) = (xs_i)$ for all $x \in M$, and therefore $Hom_R(M, M^{(I)})$ is a free left S-module.

THEOREM 5. Let M be a finitely generated left R-module with endomorphism ring S. Then the following conditions are equivalent:

(1) For every set I, every locally split submodule of $M^{(I)}$ is a direct summand of $M^{(I)}$.

(2) Every locally split submodule of $M^{(N)}$ is a direct summand of $M^{(N)}$, where N is the set of all natural numbers.

(3) Every locally split submodule of $M^{(N)}$ which is isomorphic to $M^{(N)}$ is a direct summand of $M^{(N)}$.

(4) S is a left perfect ring.

Proof. The implications (1) $\Rightarrow$ (2) and (2) $\Rightarrow$ (3) are clear. In order to prove (3) $\Rightarrow$ (4), let $U = M^{(N)}$ and $V = Hom_R(M, U)$. Then the left S-module V is isomorphic to $S^{(N)}$, i.e., V is a free left S-module with an infinitely countable number of free basis elements, say $v_1, v_2, v_3, \cdots$. Let $a_1, a_2, a_3, \cdots$ by any infinite sequence in S, and let $w_1 = v_1 - a_1 v_2$, $w_2 = v_2 - a_2 v_3$, $w_3 = v_3 - a_3 v_4, \cdots$. Let V_0 be the submodule of V generated by these elements $w_1, w_2, w_3, \cdots$. Let n and m be positive integers such that $n < m$. Then $w_1, w_2, \cdots, w_n, v_{n+1}, \cdots, v_m$ are linearly independent (over S), because for any elements $s_1, s_2, \cdots, s_m$ of S we have

$$s_1 w_1 + s_2 w_2 + \cdots + s_n w_n + s_{n+1} v_{n+1} + \cdots + s_m v_m$$

$$= s_1 v_2 + (s_2 - s_1 a_1) v_2 + \cdots + (s_n - s_{n-1} a_{n-1}) v_n$$

$$+ (s_{n+1} - s_n a_n) v_{n+1} + s_{n+2} v_{n+2} + \cdots + s_m v_m$$

and $v_1, v_2, \cdots, v_m$ are linearly independent. This implies that the sequence $w_1, w_2, \cdots, w_n, v_{n+1}, v_{n+2}, \cdots$ is linearly independent and in particular $w_1, w_2, \cdots, w_n$

are linearly independent for every n, but the latter fact is equivalent to saying that the sequence $w_1, w_2, w_3, \cdots$ forms a free basis of V_0. On the other hand, if $k \leq n$ then we have (Cf. [1], p. 313)

$$v_k = w_k + a_k w_{k+1} + \cdots + (a_k \cdots a_{n-1})w_n + (a_k \cdots a_n)v_{n+1} .$$

This shows that the sequence $w_1, w_2, \cdots, w_n, v_{n+1}, v_{n+2}, \cdots$ forms a free basis of V and, therefore, the finite direct sum $Sw_1 \oplus Sw_2 \oplus \cdots \oplus Sw_n$ is a direct summand of V. Since this is true for every n, we know that $V_0 = Sw_1 \oplus Sw_2 \oplus Sw_3 \oplus \cdots$ is locally split in V. Now let $U_0 = M \otimes_S V_0$. Then, by Theorem 2, U_0 is a locally split submodule of U. Moreover, since V_0 is isomorphic to V (by associating each w_n with v_n), U_0 is necessarily isomorphic to $M \otimes_S V = U = M^{(N)}$. Assume now the condition (3). Then it follows that U_0 is a direct summand of U. Since however $V_0 = Hom_R(M, U_0)$ by Theorem 2, V_0 becomes a direct summand of $V = Hom_R(M, U)$ (because if $U = U_0 \oplus U_1$ with a submodule U_1 of U then it follows $Hom_R(M, U) = Hom_R(M, U_0) \oplus Hom_R(M, U_1)$). According to Bass [4], Lemma 1.3 (or [1], Lemma 28.2), that V_0 is a direct summand of V implies that the descending chain $a_1 S \supset a_1 a_2 S \supset a_1 a_2 a_3 S \supset \cdots$ of principal right ideals of S terminates. Since this is true for every sequence $a_1, a_2, \cdots$ in S, this means the descending chain condition for principal right ideals of S. But this condition is equivalent to the condition that S is left perfect by [4], Theorem P. Thus (3) $\Rightarrow$ (4) is proved.

Assume conversely that S is left perfect. Let I be a set, and let U_0 be a locally split submodule of $U = M^{(I)}$. Since U is M-coreflexive, i.e., $U = M \otimes_S V$ for $V = Hom_R(M, U)$, we know by Theorem 2 that $U_0 = M \otimes_S V_0$ for $V_0 = Hom_R(M, U_0)$ and V_0 is a locally split and hence pure submodule of V. Moreover, $V = Hom_R(M, M^{(I)})$ is a free left S-module and, therefore, the factor module V/V_0 is flat as is well known. The left perfectness of S then implies that V/V_0 is projective by [4], Theorem P or [1], Theorem 28.4. Thus V_0 is a direct summand of V. But this implies that $U_0 = M \otimes_S V_0$ is a direct summand of $U = M \otimes_S V$; indeed, if $V = V_0 \oplus V_1$ with a submodule V_1 of V then we have $M \otimes_S V = M \otimes_S V \oplus M \otimes_S V_1$. Since $U = M^{(I)}$, this proves (4) $\Rightarrow$ (1).

Finally, in connection with Theorem 5, we give the following proposition on those modules whose locally split submodules are direct summands, which is more or less known for the pure submodules case:

PROPOSITION 6. Let M be a module in which every locally split submodule is a direct summand of M. Then

(i) M is a direct sum of indecomposable submodules.

(ii) If M is a direct sum of (indecomposable) submodules with local endomorphism rings, then the decomposition complements every direct summand of M.

Proof. Suppose there is given an (not necessarily countable) ascending chain of direct summands of M. Then it is easy to see that the union of these direct summands is a locally split submodule and hence a direct summand of M. Therefore, (i) can be regarded as a particular case of Stenström [6], Theorem 1. In order to prove (ii), let

$M = \oplus_I U_i$ be a direct decomposition into submodules U_i $(i \in I)$ w endomorphism rings. Let N be a proper direct summand of M, and let $M = N \oplus N'$ with a submodule $N' \neq 0$ of M. As can be seen easily, every locally split submodule of the direct summand N' of M is locally split in M and therefore is a direct summand of M whence of N'. Thus, by (i), N' is a direct sum $N' = \oplus V_j$ of indecomposable submodules V_j. Thus we have the direct decomposition $M = N \oplus (\oplus V_j)$. Choose an index j_0. Let $p_0 : M \to V_{j_0}$ be the projection with respect to the decomposition. Since the endomorphism ring of every U_i is a local ring, it follows from Azumaya [2], Theorem 1 that there exists an index $i_0 \in I$ such that the restriction of p_0 to U_{i_0} is an isomorphism $U_{i_0} \to V_{j_0}$. Since the kernel of p_0 is $N \oplus (\oplus_{j \neq j_0} V_j)$, we have that $M = U_{i_0} \oplus N \oplus (\oplus_{j \neq j_0} V_j)$ and so $N \oplus U_{i_0}$ is a direct summand of M. Let I' be a subset of I. Then we define $s(I') = N \oplus (\oplus_{I'} U_i)$. Consider now the family of those subsets I' of I which satisfy (1) $N \cap (\oplus_{I'} U_i) = 0$ and (2) $s(I') = N \oplus (\oplus_{I'} U_i)$ is a direct summand of M; such a family is non-empty as seen above. Let $\{I_\alpha\}$ be any ascending chain in the family, and let I' be the union of the chain. Then clearly I' satisfies the condition (1) and $s(I')$ is the union of the ascending chain $\{s(I_\alpha)\}$ of direct summands of M. Therefore, $s(I')$ is also a direct summand of M, *i.e.*, I' satisfies the condition (2) too, which means that I' is in the family. Thus, by Zorn's lemma, there exists a maximal member, say I_0 in the family. $s(I_0) = N \oplus (\oplus_{I_0} U_i)$ is then a direct summand of M. Suppose this were a proper submodule of M. Then by applying the above argument to $s(I_0)$ instead of N, we know the existence of an index $i_0 \in I$ such that $s(I_0) \cap U_{i_0} = 0$, whence $i_0 \notin I_0$ and $s(I_0) \oplus U_{i_0}$ is a direct summand of M. But this contradicts to the maximality of I_0, because if we put $I' = I_0 \cup \{i_0\}$ in this case I' clearly satisfies the above conditions (1) and (2). Thus we have $M = s(I_0) = M \oplus (\oplus_{I_0} U_i)$, which shows that the direct decomposition $M = \oplus_I U_i$ complements the direct summand N.

REFERENCES

1. F.W. Anderson and K.R. Fuller, Rings and categories of modules, Springer-Verlag, New York-Berlin, 1974.

2. G. Azumaya, Corrections and supplementaries to my paper concerning Krull-Remak-Schmidt's theorem, Nagoya Math. J. 1 (1950) 117-124.

3. G. Azumaya, Some characterizations of regular modules, to appear in Publicacions Matemàtiques.

4. H. Bass, Finitistic dimension and a homological generalization of semi-primary rings, Trans. Amer. Math. Soc. 95 (1960) 466-488.

5. V.S. Ramamurthi and K.M. Rangaswamy, On finitely injective modules, J. Austral. Math. Soc. 16 (1973) 239-248.

6. B. Stenström, Direct sum decompositions in Grothendieck categories, Arkiv Math. 7 (1968) 427-432.

Department of Mathematics
Indiana University
Bloomington, Indiana 47405

MODULES WITH REGULAR, PERFECT, NOETHERIAN OR ARTINIAN ENDOMORPHISM RINGS

Soumaya Makdissi Khuri

1. INTRODUCTION

The problem of classifying properties of the endomorphism ring of a module in terms of properties of the module is most naturally approached via a bijection between certain left or right ideals of the endomorphism ring and certain submodules of the module. In this paper, we use this approach to investigate when the endomorphism ring of a module is (von Neumann) regular, left or right perfect, Noetherian or artinian.

Throughout this paper, let $_R M$ be a left R-module, where R is an associative ring with identity, and let $B = End_R M$ be the ring of R-endomorphisms of M. Let

$$K = \{U \subseteq M : U = Ker\, b,\ b \in B\}\,,$$

$$I = \{U \subseteq M : U = Im\, b,\ b \in B\}\,,$$

$$T = \{U \subseteq M : U \text{ is a direct summand in } M\}\,.$$

For a general module M, it is known that B is a regular ring if and only if $K \subseteq T$ and $I \subseteq T$; for particular modules, regularity of B can be determined by either K or I, for example, for M a free module, B is regular if and only if $I \subseteq T$. One question we try to resolve in this paper is the question of when K of I determines regularity of B, and one way to do this is to determine for which modules there is a bijection, b1, between the principal right ideals of B and the submodules in K, and for which modules there is a bijection, b2, between the principal left ideals of B and the submodules in I (Proposition 2.3). Given the bijection b1 (resp. b2), we are able to deduce that regularity of B is determined by K (resp. I) by making use of a basic theorem (Theorem 2.4) which allows us to transfer the property of being a direct summand from B to M and conversely.

Since the definition of a regular ring is right-left symmetric, it is natural to ask here how the conditions "$K \subseteq T$" and "$I \subseteq T$" are related to each other and to the existence of the bijections b1 and b2. In answer, we find that "$I \subseteq T$" implies the existence of the bijection b1, while "$K \subseteq T$" implies the existence of the bijection b2 (Theorem 3.3). Moreover, we have (Theorem 3.5):

(i) Given $K \subseteq T$, then $I \subseteq T$ if and only if for each $U \in T$, and each monomorphism $\sigma : U \to M$, $Im\, \sigma \in T$.

(ii) Given $I \subseteq T$, then $K \subseteq T$ if and only if, for each $U \in T$, and each epimorphism $\sigma : M \to U$, $Ker\, \sigma \in T$.

An easy application of Theorem 3.5 gives that, for M continuous, regularity of B is equivalent to $K \subseteq T$, while for M nonsingular and CS, regularity of B is equivalent to $I \subseteq T$; for M nonsingular and continuous, B is regular (Proposition 3.6).

The bijections b1 and b2 also serve to determine when B is left or right perfect (Proposition 4.1). In order to determine when B is left or right Noetherian or artinian, similar bijections can be established between the finitely generated left or right ideals of B and certain classes, $\mathcal{M}^a$ or $\mathcal{M}^\bullet$, of submodules of M. We find, for example, that, in the presence of a bijection between the right ideals of B and the submodules in $\mathcal{M}^a$, we have that B is right Noetherian if and only if M satisfies the a.c.c. on $\mathcal{M}^a$ (Theorem 4.3 (i)) and B is right artinian if and only if M satisfies the a.c.c. and the d.c.c. on $\mathcal{M}^a$ (Theorem 4.6 (i)).

2. CORRESPONDENCE THEOREMS

In this section, we establish the bijections needed to determine necessary and sufficient conditions on M in order that B should be regular, left or right perfect, Noetherian or artinian. Recall that B is a *regular* ring if and only if every principal right (and left) ideal of B is a direct summand in B; and that B is a left (resp. right) *perfect* ring if and only if B satisfies the descending chain condition (d.c.c.) on principal right (resp. left) ideals. Hence, when investigating whether B is regular or perfect, we need to establish bijections between the principal left or right ideals of B and certain submodules of M.

Set

$$\mathcal{P}^r(B) = \{K \subseteq B : K = bB, b \in B\}$$

and

$$\mathcal{P}^\ell(B) = \{H \subseteq B : H = Bb, b \in B\} .$$

The needed bijections will be induced by the following well-known maps: For a submodule, U, of M, and a subset, H, of B, define:

$$r_B(U) = \{b \in B : Ub = 0\} , \; l_M(H) = \{m \in M : mH = 0\} ;$$
$$I_B(U) = \{b \in B : Mb \subseteq U\} , \; S_M(H) = MH = \sum_{h \in H} Mh.$$

Clearly, $r_B(U)$ is a right ideal of B, while $I_B(U)$ is a left ideal of B, and $l_M(H)$ and $S_M(H)$ are submodules of M. The notations $S_M(H)$ and MH will be used interchangeably, and we will identify $I_B(U)$ and $Hom_R(M, U)$.

The following Propositions are easily verified:

Proposition 2.1: Let U_1 and U_2 be submodules of M, H_1 and H_2 be subsets of B; then:

(i) $U_1 \subseteq U_2 \Rightarrow r_B(U_1) \supseteq r_B(U_2)$.
(ii) $H_1 \subseteq H_2 \Rightarrow l_M(H_1) \supseteq l_M(H_2)$.
(iii) $U_1 \subseteq l_M r_B(U_1)$ and $H_1 \subseteq r_B l_M(H_1)$.
(iv) $r_B(U_1) = r_B l_M r_B(U_1)$ and $l_M(H_1) = l_M r_B l_M(H_1)$.

Proposition 2.2: For U_1, U_2 submodules of M, H_1, H_2 subsets of B, we have:

(i) $U_1 \subseteq U_2 \Rightarrow I_B(U_1) \subseteq I_B(U_2)$.

(ii) $H_1 \subseteq H_2 \Rightarrow S_M(H_1) \subseteq S_M(H_2)$.

(iii) $S_M I_B(U_1) \subseteq U_1$ and $H_1 \subseteq I_B S_M(H_1)$.

(iv) $I_B(U_I) = I_B S_M I_B(U_1)$ and $S_M(H_1) = S_M I_B S_M(H_1)$.

It follows from Propositions 2.1 and 2.2 that r_B and l_M induce order-reversing maps between the submodules of M and the right ideals of B, while I_B and S_M induce order-preserving maps between the submodules of M and the left ideals of B.

Recall that an order-preserving (resp. order-reversing) bijection between two partially ordered sets is called a *projectivity* (resp. a *duality*).

Set $\mathcal{M}^a = \{U \subseteq M : U = l_M r_B(U)\} = \{U \subseteq M : U = l_M(K), \text{ for some } K \subset B\}$,

and

$$\mathcal{B}^a = \{K \subseteq B : K = r_B l_M(K)\} = \{K \subseteq B : K = r_B(U), \text{ for some } U \subseteq M\}.$$

Clearly, for $b \in B$, $l_M(b) = Ker\, b$, so $K \subseteq \mathcal{M}^a$.

Set $\mathcal{M}^{\cdot} = \{U \subseteq M : U = S_M I_B(U)\} = \{U \subseteq M : U = S_M(H), \text{ for some } H \subseteq B\}$,

and

$$\mathcal{B}^{\cdot} = \{H \subseteq B : H = I_B S_M(H)\} = \{H \subseteq B : H = I_B(U), \text{ for some } U \subset M\}.$$

Clearly, for $b \in B$, $S_M(b) = Mb = Im\, b$, so $I \subseteq \mathcal{M}^{\cdot}$.

Since $l_M(bB) = l_M(b)$ and $S_M(Bb) = S_M(b)$, it is clear that l_M maps $P^r(B)$ to K and S_M maps $P^l(B)$ to I. Moreover, the conditions which are necessary and sufficient in order for these maps to be bijections are the obvious ones, as the next proposition states:

Proposition 2.3:

(i) The maps r_B and l_M determine a duality between K and $P^r(B)$ if and only if, for each $b \in B$, $bB = r_B l_M(bB)$ (i.e., if and only if $P^r(B) \subseteq \mathcal{B}^a$).

(ii) The maps I_B and S_M determine a projectivity between I and $P^l(B)$ if and only if, for each $b \in B$, $Bb = I_B S_M(Bb)$ (i.e., if and only if $P^l(B) \subseteq \mathcal{B}^{\cdot}$).

Given one of the bijections in Proposition 2.3, say, for example, the duality in (i), it is clear that the d.c.c. on $P^r(B)$ is equivalent to the a.c.c. on K. It is less obvious, but nevertheless true, that every element of $P^r(B)$ is a direct summand in B is and only if every element of K is a direct summand in M. This follows immediately from the following general result.

Theorem 2.4:

(i) If r_B and l_M determine a duality between $\mathcal{U} \subseteq \mathcal{M}^a$ and $\mathcal{V} \subseteq \mathcal{B}^a$, then every $K \in \mathcal{V}$ is a direct summand in B if and only if every $U \in \mathcal{U}$ is a direct summand in M.

(ii) If I_B and S_M determine a projectivity between $\mathcal{U} \subseteq M^\bullet$ and $\mathcal{V} \subseteq \mathcal{B}^\bullet$, then every $K \in \mathcal{V}$ is a direct summand in B if and only if every $U \in \mathcal{U}$ is a direct summand in M.

Before proving Theorem 2.4, we need a small Lemma and some notation for annihilators in B : For $H \subseteq B$, let

$$\mathcal{L}(H) = \{b \in B : bH = 0\} \text{ and } \mathcal{R}(H) = \{b \in B : Hb = 0\} .$$

Lemma 2.5: For $H \subseteq B$, $I_B l_M (H) = \mathcal{L}(H)$ and $r_B S_M (H) = \mathcal{R}(H)$.

Proof: $b \in I_B l_M (H) \Leftrightarrow Mb \subseteq l_M (H) \Leftrightarrow (Mb)H = 0 \Leftrightarrow bH = 0 \Leftrightarrow b \in \mathcal{L}(H)$. $b \in r_B S_M (H) \Leftrightarrow [S_M (H)]b = 0 \Leftrightarrow MHb = 0 \Leftrightarrow Hb = 0 \Leftrightarrow b \in \mathcal{R}(H)$.

Proof of Theorem 2.4: (i) Given that r_B and l_M determine a duality between $\mathcal{U} \subseteq M^a$ and $\mathcal{V} \subseteq \mathcal{B}^a$, assume that every $U \in \mathcal{U}$ is a direct summand in M and let $K \in \mathcal{V}$. Then $U = l_M (K) \in \mathcal{U}$, hence there is an idempotent, $e = e^2 \in B$, such that $U = Me$. Since $K \in \mathcal{V} \subseteq \mathcal{B}^a$, we have $K = r_B l_M (K)$; then, using Lemma 2.5, we have $K = r_B l_M (K) = r_B (Me) = \mathcal{R}(e) = (1 - e)B$, i.e., K is a direct summand in B.

Conversely, assume that every $K \in \mathcal{V}$ is a direct summand in B, and let $U \in \mathcal{U}$. Then $K = r_B (U) \in \mathcal{V}$, hence there is an idempotent, $e = e^2 \in B$, such that $K = eB$. Then, since $U \in \mathcal{U} \subseteq M^a$, we have $U = l_M r_B (U) = l_M (eB) = l_M (e) = M(1 - e)$, i.e., U is a direct summand in M.

(ii) Given that I_B and S_M determine a projectivity between $\mathcal{U} \subseteq M^\bullet$ and $\mathcal{V} \subseteq \mathcal{B}^\bullet$, assume that every $U \in \mathcal{U}$ is a direct summand in M, and let $H \in \mathcal{V}$. Then $U = S_M (H) \in \mathcal{U}$, hence $U = Me$, for $e = e^2 \in B$. Then $H = I_B S_M (H) = I_B (Me)$. We will show that $I_B (Me) = Be$ when e is an idempotent, from which it will follow that H is a direct summand in B.

Clearly, $e \in I_B (Me)$, hence $Be \subseteq I_B (Me)$. On the other hand, let $b \in I_B (Me)$; then, for any $m \in M$, $mb = m_1 e$ for some $m_1 \in M$, hence $mbe = m_1 e^2 = m_1 e = mb$, i.e., $b = be \in Be$.

Conversely, assume that every $H \in \mathcal{V}$ is a direct summand in B and let $U \in \mathcal{U}$. Then $H = I_B (U) \in \mathcal{V}$, hence $H = Be$, for $e = e^2 \in B$. Then $U = S_M I_B (U) = S_M (Be) = Me$, i.e., U is a direct summand in M.

An equivalent definition of regularity of B is that every finitely generated $(f.g.)$ left (and right) ideal of B should be a direct summand in B. Moreover, recall that B is right (left) Noetherian if and only if B satisfies the $a.c.c.$ on $f.g.$ right (left) ideals. Thus it is of interest to us to establish correspondence theorems for $f.g.$ right or left ideals.

Set $\mathcal{F}^r (B) = \{K \subseteq B : K = \sum_{i=1}^n b_i B, b_i \in B\}$ and

$$\mathcal{F}^l (B) = \{H \subseteq B : H = \sum_{i=1}^n Bb_i , b_i \in B\} .$$

Also set $M_f^a = \{U \subseteq M : U = \bigcap_{i=1}^{n} l_M(b_i), \, b_i \in B\}$

and

$$M_f^{\bullet} = \{U \subseteq M : U = \sum_{i=1}^{n} Mb_i, \, b_i \in B\} \, .$$

Note that $\bigcap_{i=1}^{n} l_M(b_i) = l_M\left(\sum_{i=1}^{n} b_i B\right) \in M^a$, i.e., $M_f^a \subseteq M^a$,

and

$$\sum_{i=1}^{n} Mb_i = M \sum_{i=1}^{n} Bb_i = S_M\left(\sum_{i=1}^{n} Bb_i\right) \in M^{\bullet}, \text{ i.e., } M_f^{\bullet} \subseteq M^{\bullet} \, .$$

Proposition 2.6:

(i) The maps r_B and l_M determine a duality between M_f^a and $\mathcal{F}^r(B)$ if and only if $K = r_B l_M(K)$ for each $K \in \mathcal{F}^r(B)$ (i.e., if and only if $\mathcal{F}^r(B) \subseteq B^a$).

(ii) The maps I_B and S_M determine a projectivity between $M_f^{\bullet}$ and $\mathcal{F}^l(B)$ if and only if $H = I_B S_M(H)$ for each $H \in \mathcal{F}^l(B)$ (i.e., if and only if $\mathcal{F}^l(B) \subseteq B^{\bullet}$).

Proof: (i) One direction is obvious. Conversely, if $K = r_B l_M(K)$ for each $K \in \mathcal{F}^r(B)$, then we have, for $U = \bigcap_{i=1}^{n} l_M(b_i) \in M_f^a$ and for $K = \sum_{i=1}^{n} b_i B \in \mathcal{F}^r(B)$:

$$U \rightarrow r_B(U) = r_B l_M\left(\sum_{i=1}^{n} b_i B\right) = \sum_{i=1}^{n} b_i B \in \mathcal{F}^r(B) \rightarrow l_M r_B(U) = U$$

and

$$K \rightarrow l_M(K) = l_M\left(\sum_{i=1}^{n} b_i B\right) = \bigcap_{i=1}^{n} l_M(b_i) \in M_f^a \rightarrow r_B l_M(K) = K \, .$$

Hence, r_B and l_M determine a duality between M_f^a and $\mathcal{F}^r(B)$.

(ii) Here again, one direction is obvious. Conversely, if $H = I_B S_M(H)$ for each $H \in \mathcal{F}^l(B)$, then we have for $U = \sum_{i=1}^{n} Mb_i \in M_f^{\bullet}$ and for $H = \sum_{i=1}^{n} Bb_i \in \mathcal{F}^l(B)$:

$$U \rightarrow I_B\left(\sum_{i=1}^{n} Mb_i\right) = I_B S_M\left(\sum_{i=1}^{n} Bb_i\right) = \sum_{i=1}^{n} Bb_i \in \mathcal{F}^l(B) \rightarrow S_M I_B(U) = U;$$

and

$$H \rightarrow S_M\left(\sum_{i=1}^{n} Bb_i\right) = \sum_{i=1}^{n} Mb_i \in M_f^{\bullet} \rightarrow I_B S_M(H) = H \, .$$

Hence, I_B and S_M determine a projectivity between M_r^* and $\mathcal{F}^l(B)$.

3. REGULAR ENDOMORPHISM RINGS

A combination of Theorem 2.4 and Proposition 2.3 gives dual conditions on M in order for K or I to determine regularity of B :

Theorem 3.1:

(i) If $P^r(B) \subseteq B^a$, then B is a regular ring if and only if $K \subseteq T$.

(ii) If $P^l(B) \subseteq B^*$, then B is a regular ring if and only if $I \subseteq T$.

Proof: (i) If $P^r(B) \subseteq B^a$, then by Proposition 2.3, r_B and l_M determine a duality between K and $P^r(B)$. Since, as we have noted, $K \subseteq M^a$, and, by hypothesis $P^r(B) \subseteq B^a$, it follows by Theorem 2.4 that every $K \in P^r(B)$ is a direct summand in B if and only if every $U \in K$ is a direct summand in M; in other words, B is regular if and only if $K \subseteq T$.

The proof of (ii) is analogous to the proof of (i).

Corollary 3.2:

(i) If $_R M$ is quasi-injective, then B is regular if and only if $K \subseteq T$.

(ii) If $_R M$ is quasi-projective, then B is regular if and only if $I \subseteq T$.

Proof: It is known that, when M is quasi-injective (resp. quasi-projective), we have $K = r_B l_M(K)$ (resp. $H = I_B S_M(H)$) for every finitely generated right (resp. left) ideal of $B(cf\ e.g.,$ [1, Propositions 4.1 and 4.9]).

In Theorem 3.1, we saw that the condition "$P^r(B) \subseteq B^a$" implied that regularity of B was equivalent to "$K \subseteq T$" while the condition "$P^l(B) \subseteq B^*$" implied that regularity of B was equivalent to "$I \subseteq T$". We will now show that "$I \subseteq T$" implies "$P^r(B) \subseteq B^a$" while "$K \subseteq T$" implies "$P^l(B) \subseteq B^*$".

Theorem 3.3:

(i) If $I \subseteq T$, then $P^r(B) \subseteq B^a$.

(ii) If $K \subseteq T$, then $P^l(B) \subseteq B^*$.

Proof:

(i) Assume that $I \subseteq T$, and let $K = bB \in P^r(B)$, $b \in B$. Let $c \in r_B l_M(bB) = r_B l_M(b)$, so that $[l_M(b)]c = 0$ or $l_M(b) \subseteq l_M(c)$. Define $h_1 : Mb \to Mc$ by $(mb)h_1 = mc$, for each $m \in M$. h_1 is well-defined since: $mb = m_1 b \Rightarrow m - m_1 \in l_M(b) \subseteq l_M(c) \Rightarrow mc = m_1 c$. Since $Mb \in T$, we have $M = Mb \oplus U$ for some $U \subseteq M$, so we can extend h_1 to $h \in B$ by defining $uh = 0$ for $u \in U$. Then, for each $m \in M$, we have $(mb)h = (mb)h_1 = mc$, i.e., $c = bh \in bB$. Thus $r_B l_M(bB) \subseteq bB$, and since, always, $bB \subseteq r_B l_M(bB)$ (Proposition 2.1), it follows that $bB = r_B l_M(bB)$, i.e., $K \in B^a$.

(ii) Assume that $K \subseteq T$ and let $b \in B$. Always, $Bb \subseteq I_B S_M (Bb)$ (Proposition 2.2); conversely, let $c \in I_B S_M (Bb) = I_B S_M (b)$, so that $Mc \subseteq S_M (b) = Mb$. Since $l_M (b)$ is a direct summand in M, there is $e = e^2 \in B$ such that $l_M (b) = l_M (e)$ and $M = l_M (b) \oplus Me = l_M (e) \oplus Me$. Also, $l_M (c) \in T$, so there is $f = f^2 \in B$ such that $l_M (c) = l_M (f)$ and $M = l_M (c) \oplus Mf = l_M (f) \oplus Mf$.

Given $mf \in Mf$, we have $mc \in Mc \subseteq Mb$, hence, there is $m_1 \in M$ such that $mc = m_1 b$. Define $h_1 : Mf \to Me$ by $(mf) h_1 = m_1 e$, where $mc = m_1 b$. Then h_1 is well-defined since: $mf = m'f \Rightarrow m - m' \in l_M (f) = l_M (c) \Rightarrow mc = m'c$, and if $m_1 b = mc = m_1' b$, then $m_1 - m_1' \in l_M (b) = l_M (e)$, so $m_1 e = m_1' e$. Since $Mf \in T$, h_1 extends to $h \in B$ by defining $xh = 0$ for $x \in l_M (f)$. Note that, for any $m \in M$, $m = m(1 - e) + me$, and $m(1 - e) \in l_M (e) = l_M (b)$, so that $mb = meb$. Now, for any $m \in M$, we have $m = mf + m(1 - f)$, and $mh = mfh = m_1 e$ where $m_1 b = mc$; then $mhb = m_1 eb = m_1 b = mc$, so that $c = hb \in Bb$, and $Bb = I_B S_M (Bb)$.

It is interesting to obtain the following known result via Theorems 3.1 and 3.3.

Corollary 3.4: For any ${}_R M$, B is regular if and only if $K \subseteq T$ and $I \subseteq T$.

Proof: If B is regular, then for any $b \in B$, there is $e = e^2 \in B$ such that $bB = eB$, and there is $f = f^2 \in B$ such that $Bb = Bf$. Then $l_M (b) = l_M (bB) = l_M (eB) = l_M (e)$, so $l_M (b) \in T$; and $S_M (b) = S_M (Bb) = S_M (Bf) = Mf$, so $S_M (b) \in T$.

Conversely, assume that $K \subseteq T$ and $I \subseteq T$. By Theorem 3.3(i), since $I \subseteq T$, we have $P^r (B) \subseteq B^a$, hence, by Theorem 3.1(i), B is regular since $K \subseteq T$. Note that we could just as easily have combined 3.3(ii) and 3.1(i) to prove B regular.

Our next result gives a condition on M equivalent to one of the conditions "$K \subseteq T$" or "$I \subseteq T$" in case the other one is given.

Theorem 3.5:

(i) Assume that $K \subseteq T$. Then $I \subseteq T$ if and only if, for each $U \in T$ and each monomorphism $\sigma : U \to M$, $U \sigma \in T$.

(ii) Assume that $I \subseteq T$. Then $K \subseteq T$ if and only if, for each $U \in T$ and each epimorphism $\sigma : M \to U$, $l_M (\sigma) \in T$.

Proof: (i) Assume that $K \subseteq T$. Suppose that also $I \subseteq T$ and let $\sigma : U \to M$ be a monomorphism, where $U \in T$. We have $M = U \oplus V$, for some $V \subseteq M$; define $b \in B$ by: $ub = u\sigma$ for $u \in U$ and $vb = 0$ for $v \in V$. Then $Mb = Ub = U\sigma$, hence since $Mb \in T$, $U\sigma \in T$.

Conversely, suppose that whenever $\sigma : U \to M$ is a monomorphism and $U \in T$ then $U\sigma \in T$. Let $b \in B$; by hypothesis, we have $M = l_M (b) \oplus V$ for some $V \subseteq M$. Let σ be the restriction of b to V; then $\sigma : V \to M$ is a monomorphism and $V \in T$, hence $Mb = Vb = V\sigma \in T$.

(ii) Assume that $I \subseteq T$. Suppose that also $K \subseteq T$ and let $\sigma : M \to U$ be an epimorphism, where $U \in T$. Then $\sigma \in B$, so $l_M (\sigma) \in T$.

Conversely, suppose that whenever $\sigma : M \to U$ is an epimorphism and $U \in \mathcal{T}$, then $l_M(\sigma) \in \mathcal{T}$. Let $b \in B$; by hypothesis, we have $M = Mb \oplus N$, for some $N \subseteq M$, i.e., $b : M \to Mb$ is an epimorphism with $Mb \in \mathcal{T}$. Hence $l_M(b) \in \mathcal{T}$, i.e., $K \subseteq \mathcal{T}$.

As an application of Theorem 3.5, we will determine when the endomorphism ring of a CS module or a continuous module is regular. Recall that M is a *CS module* if every complement submodule of M is a direct summand in M; M is a *continuous module* if (a) M is a CS module and (b) the isomorphic image of a direct summand in M is a direct summand in M.

Proposition 3.6:

(i) If M is continuous, then B is regular if and only if $K \subseteq \mathcal{T}$.

(ii) If M is nonsingular and CS, then B is regular if and only if $I \subseteq \mathcal{T}$.

(iii) If M is nonsingular and continuous, then B is regular.

Proof:

(i) Assume that M is continous. If B is regular, then, by Corollary 3.4, $K \subseteq \mathcal{T}$. Conversely, suppose that $K \subseteq \mathcal{T}$; then, for any $U \in \mathcal{T}$ and any monomorphism $\sigma : U \to M$, we have $U\sigma \in \mathcal{T}$ since M is continous. Hence, by Theorem 3.5, $I \subseteq \mathcal{T}$ and, therefore, by Corollary 3.4, B is regular.

(ii) Assume that M is nonsingular and CS. Since M is nonsingular, every $U \in K$ is a complement (cf e.g., [4, Lemma 3.12 (proof)]). Hence, since M is also CS, every $U \in K$ is a direct summand in M. It follows by Corollary 3.4 that B is regular if and only if $I \subseteq \mathcal{T}$.

(iii) Assume that M is nonsingular and continuous. Then M is, in particular, nonsingular and CS, hence, as we have just seen in (ii), $K \subseteq I$. It now follows by (i) that B is regular.

Corollary 3.7: If M is nonsingular and continuous, then, for any $b \in B$, we have $bB = r_B l_M(bB)$ and $Bb = I_B S_M(Bb)$.

Proof: If M is nonsingular and continuous, then B is regular by Proposition 3.6, hence $K \subseteq \mathcal{T}$ and $I \subseteq \mathcal{T}$ by Corollary 3.4, hence $P^r(B) \subseteq B^a$ and $P^l(B) \subseteq B^s$ by Theorem 3.3.

4. PERFECT, SEMIPRIMARY, NOETHERIAN AND ARTINIAN ENDOMORPHISM RINGS

Recall that B is a left (right) perfect ring if and only if B satisfies the *d.c.c.* on principal right (left) ideals, if and only if B satisfies the *d.c.c.* on finitely generated right (left) ideals.

Thus, we can deduce from Proposition 2.3:

Proposition 4.1:

(i) If $P^r(B) \subseteq B^a$, then B is a left perfect ring if and only if M satisfies the *a.c.c.* on K.

(ii) If $P^l(B) \subseteq B^\bullet$, then B is a right perfect ring if and only if M satisfies the *d.c.c.* on I.

In particular, in view of Corollary 3.7, if M is nonsingular and continuous, then B is left (right) perfect if and only if M satisfies the *a.c.c.* (*d.c.c.*) on $K(I)$.

From Proposition 2.6, we can deduce:

Proposition 4.2:

(i) If $\mathcal{F}^r(B) \subseteq B^a$, then: B is regular if and only if $M_f^a \subseteq \mathcal{T}$ and B is left perfect if and only if M satisfies the *a.c.c.* on M_f^a.

(ii) If $\mathcal{F}^l(B) \subseteq B^\bullet$, then: B is regular if and only if $M_f^\bullet \subseteq \mathcal{T}$ and B is right perfect if and only if M satisfies the *d.c.c.* on $M_f^\bullet$.

Proposition 2.6 is also useful in determining when B is Noetherian:

Theorem 4.3:

(i) If $\mathcal{F}^r(B) \subseteq B^a$, then B is right Noetherian if and only if M satisfies the *d.c.c.* on M_f^a, if and only if M satisfies the *d.c.c.* on M^a.

(ii) If $\mathcal{F}^l(B) \subseteq B^\bullet$, then B is left Noetherian if and only if M satisfies the *a.c.c.* on $M_f^\bullet$, if and only if M satisfies the *a.c.c.* on $M^\bullet$.

Proof:

(i) The first equivalence is clear if we make use of the duality given by Proposition 2.6(i).

To see that the second equivalence holds, assume that B is right Noetherian and let $U_1 \supseteq U_2 \supseteq \cdots$ be a descending chain of submodules in M^a. Then $r_B(U_1) \subseteq r_B(U_2) \subseteq \cdots$ is an ascending chain of right ideals in B, hence there is an integer $n > 0$ such that $r_B(U_{n+j}) = r_B(U_n)$ for $j \geq 1$, and this implies that $U_{n+j} = l_M r_B(U_{n+j}) = l_M r_B(U_n) = U_n$ for $j \geq 1$, *i.e.*, M satisfies the *d.c.c.* on M^a. Conversely, if M satisfies the *d.c.c.* on M^a, then it satisfies the *d.c.c.* on M_f^a, and B is right Noetherian.

(ii) Here again the first equivalence is clear by Proposition 2.6(ii). Assume that B is left Noetherian and that $U_1 \subseteq U_2 \subseteq \cdots$ is an ascending chain of submodules in $M^\bullet$. Then $I_B(U_1) \subseteq I_B(U_2) \subseteq \cdots$ is an ascending chain of left ideals in B, so there is an integer $n > 0$ such that $I_B(U_{n+j}) = I_B(U_n)$ for $j \geq 1$, hence we have $U_{n+j} = S_M I_B(U_{n+j}) = S_M I_B(U_n) = U_n$, for $j \geq 1$, *i.e.*, M satisfies the *a.c.c.* on $M^\bullet$. Conversely, if M satisfies the *a.c.c.* on $M^\bullet$, then it satisfies the *a.c.c.* on $M_f^\bullet$, and B is left Noetherian.

In order to determine when B is artinian, we shall first determine when B is semiprimary. Recall that a ring R with Jacobson radical J is *semiprimary* if and only if R/J is semisimple and J is nilpotent.

We shall be making use of the following result:

Proposition 4.4: Let R be a ring which satisfies the *a.c.c.* on annihilator left (resp. right) ideals. If R is left (right) perfect, then R is semiprimary.

Proposition 4.4 follows immediately from the following known results:

- R is left (right) perfect if and only if R/J is semisimple and J is left (right) T-nilpotent [3].
- Let R be a ring which satisfies the *a.c.c.* on annihilator left (right) ideals. If I is a left (right) T-nilpotent one-sided ideal, then I is nilpotent [2, Proposition 29.1].

Theorem 4.5:

(i) Assume that $\mathcal{F}^r(B) \subseteq B^a$ and that M satisfies the *a.c.c.* on $\mathcal{M}^a$. Then B is semiprimary.

(ii) Assume that $\mathcal{F}^l(B) \subseteq B^\bullet$ and that M satsifies the *d.c.c.* on $\mathcal{M}^\bullet$. Then B is semiprimary.

Proof:

(i) If $\mathcal{F}^r(B) \subseteq B^a$ and M satisfies the *a.c.c.* on $\mathcal{M}^a$, then, by Proposition 4.2, B is left perfect. Hence, by Proposition 4.4, in order to show that B is semiprimary, it is sufficient to show that B satisfies the *a.c.c.* on annihilator left ideals.

Let $\mathcal{L}(K_1) \subseteq \mathcal{L}(K_2) \subseteq \cdots$ be an ascending chain of annihilator left ideals of B. Then $S_M \mathcal{L}(K_1) \subseteq S_M \mathcal{L}(K_2) \subseteq \cdots$ and we have an ascending chain, $l_M r_B S_M \mathcal{L}(K_1) \subseteq l_M r_B S_M \mathcal{L}(K_2) \subseteq \cdots$ of submodules in $\mathcal{M}^a$. Hence, there is an $n > 0$ such that $l_M r_B S_M \mathcal{L}(K_{n+j}) = l_M r_B S_M \mathcal{L}(K_n)$, for $j \geq 1$ and, therefore, $r_B S_M \mathcal{L}(K_{n+j}) = r_B S_M \mathcal{L}(K_n)$, for $j \geq 1$. Since $r_B S_M(H) = \mathcal{R}(H)$, for $H \subseteq B$, by Lemma 2.5, we have $\mathcal{R} \mathcal{L}(K_{n+j}) = \mathcal{R} \mathcal{L}(K_n)$ and, hence, $\mathcal{L}(K_{n+j}) = \mathcal{L}(K_n)$, for $j \geq 1$. This proves that B satisfies the *a.c.c.* on annihilator left ideals and consequently B is semiprimary.

(ii) If $\mathcal{F}^l(B) \subseteq B^\bullet$ and M satisfies the *d.c.c.* on $\mathcal{M}^\bullet$, then, by Proposition 4.2, B is right perfect. Hence, by Proposition 4.4, in order to show that B is semiprimary, it is sufficient to show that B satisfies the *a.c.c.* on annihilator right ideals.

Let $\mathcal{R}(H_1) \subseteq \mathcal{R}(H_2) \subseteq \cdots$ be an ascending chain of annihilator right ideals of B. Then $\mathcal{L} \mathcal{R}(H_1) \supseteq \mathcal{L} \mathcal{R}(H_2) \supseteq \cdots$ and, therefore, $S_M \mathcal{L} \mathcal{R}(H_1) \supseteq S_M \mathcal{L} \mathcal{R}(H_2) \supseteq \cdots$ is a descending chain of submodules in $\mathcal{M}^\bullet$. Hence, there is an $n > 0$ such that $S_M \mathcal{L} \mathcal{R}(H_{n+j}) = S_M \mathcal{L} \mathcal{R}(H_n)$, for $j \geq 1$. By Lemma 2.5, $\mathcal{L}(H) = I_B l_M(H)$, for any $H \subseteq B$, i.e., we have $S_M I_B l_M \mathcal{R}(H_{n+j}) = S_M I_B l_M \mathcal{R}(H_n)$, for $j \geq 1$; and by Proposition 2.2, $I_B S_M I_B(U) = I_B(U)$, for any $U \subseteq M$, so we have $I_B l_M \mathcal{R}(H_{n+j}) = I_B l_M \mathcal{R}(H_n)$, for $j \geq 1$, or $\mathcal{L} \mathcal{R}(H_{n+j}) = \mathcal{L} \mathcal{R}(H_n)$ and, therefore, $\mathcal{R}(H_{n+j}) = \mathcal{R}(H_n)$, for $j \geq 1$. This proves that B satisfies the *a.c.c.* on annihilator right ideals and consequently B is semiprimary.

Now, a combination of Theorems 4.3 and 4.5 with the following well-known result will allow us to determine when B is artinian:

Hopkins' Theorem: A ring R is right (resp. left) artinian if and only if R is right (resp. left) Noetherian and semiprimary.

Theorem 4.6:

(i) Assume that $\mathcal{F}^r(B) \subseteq \mathcal{B}^a$. Then B is right artinian if and only if M satisfies the a.c.c. and the d.c.c. on $\mathcal{M}^a$.

(ii) Assume that $\mathcal{F}^l(B) \subseteq \mathcal{B}^\bullet$. Then B is left artinian if and only if M satisfies the a.c.c. and the d.c.c. on $\mathcal{M}^\bullet$.

Proof:

(i) Assume that $\mathcal{F}^r(B) \subseteq \mathcal{B}^a$, and suppose that M satisfies the a.c.c. and the d.c.c. on $\mathcal{M}^a$. Then by Theorem 4.3(i), B is right Noetherian, and by Theorem 4.5(i), B is semiprimary. Hence, by Hopkins' Theorem, B is right artinian.

Conversely, suppose that B is right artinian. Then, by Hopkins' Theorem, B is right Noetherian, and, by Theorem 4.3(i), M satisfies the d.c.c. on $\mathcal{M}^a$. Let $U_1 \subseteq U_2 \subseteq \cdots$ be an ascending chain of submodules in $\mathcal{M}^a$. Then $r_B(U_1) \supseteq I_B(U_2) \supseteq \cdots$ is a descending chain of right ideals of B, hence, by hypothesis, there is an $n > 0$ such that $r_B(U_{n+j}) = r_B(U_n)$ for $j \geq 1$. Then, $U_{n+j} = l_M\, r_B(U_{n+j}) = l_M\, r_B(U_n) = U_n$ for $j \geq 1$, i.e., M satisfies the a.c.c. on $\mathcal{M}^a$.

(ii) Assume that $F^l(B) \subseteq \mathcal{B}^\bullet$, and suppose that M satisfies the a.c.c. and the d.c.c. on $\mathcal{M}^\bullet$. Then, as in (i), by Theorem 4.3(ii), B is left Noetherian, and by Theorem 4.5(ii), B is semiprimary. Hence, by Hopkins' Theorem, B is left artinian.

Conversely, suppose that B is left artinian. Then, by Hopkins' Theorem, B is left Noetherian, and by Theorem 4.3(ii), M satisfies the a.c.c. on $\mathcal{M}^\bullet$. Let $U_1 \supseteq U_2 \supseteq \cdots$ be a descending chain of submodules in $\mathcal{M}^\bullet$. Then $I_B(U_1) \supseteq I_B(U_2) \supseteq \cdots$ is a descending chain of left ideals of B, hence, by hypothesis, there is an $n > 0$ such that $I_B(U_{n+j}) = I_B(U_n)$ for $j \geq 1$. Then, $U_{n+j} = S_M\, I_B(U_{n+j}) = S_M\, I_B(U_n) = U_n$ for $j \geq 1$, i.e., M satisfies the d.c.c. on $\mathcal{M}^\bullet$.

Corollary 4.7:

(i) Let $_RM$ be quasi-injective. Then B is right artinian if and only if M satisfies the a.c.c. and the d.c.c. on $\mathcal{M}^a$.

(ii) Let $_RM$ be quasi-projective. Then B is left artinian if and only if M satisfies the a.c.c. and the d.c.c. on $\mathcal{M}_\bullet$.

Remark: A close look at some of the preceding proofs shows that some of the "iff" statements can be proved in one direction without using the hypotheses "$\mathcal{F}^r(B) \subseteq \mathcal{B}^a$" or "$\mathcal{F}^l(B) \subseteq \mathcal{B}^\bullet$" Specifically, we have:

Proposition 4.8:

(i) If B is right Noetherian, then M satisfies the *d.c.c.* on M^a; and if B is right artinian, then M satisfies the *d.c.c.* and the *a.c.c.* on M^a.

(ii) If B is left Noetherian, then M satisfies the *a.c.c.* on $M^\bullet$; and if B is left artinian, then M satisfies the *a.c.c.* and the *d.c.c.* on $M^\bullet$.

REFERENCES

1. T. Albu and C. Nastasescu, *Relative finiteness in module theory*, Dekker, New York, 1984.
2. F.W. Anderson and K.R. Fuller, *Rings and categories of modules*, Springer-Verlag, New York, 1974.
3. H. Bass, Finitistic dimension and a homological generalization of semi-primary rings, Trans. Amer. Math. Soc. **95** (1960) 466-488.
4. S.M. Khuri, Correspondence theorems for modules and their endomorphism rings. Journal of Algebra **122** (1989) 380-396.

Mathematics Department
East Carolina University
Greenville, NC 27858

AZUMAYA RINGS AND MASCHKE'S THEOREM

Joseph A. Wehlen

The purpose of this note is to give a characterization of biregular group rings which are Azumaya rings and to extend the classical Maschke's Theorem to locally normal groups. In addition, it will provide a class of examples of Azumaya rings. A ring A with unity is said to be an Azumaya ring if and only if (i) [ideal lifting condition] every two-sided ideal of A is generated by an ideal of the center and (ii) [central epimorphism condition] the center of every homomorphic image of A is the homomorphic image of the center [Az]. Details about Azumaya rings (also known as separable rings) may be found in [Az], [Bu1], [Bu2] and [Bu3].

By a locally normal group G we shall mean that every element of the group G has finite order and is contained in a finite normal subgroup of G. A ring A is said to be biregular if every ideal of A is generated by a central idempotent. A commutative ring R shall be called locally perfect if each localization at a maximal ideal is a perfect ring. R shall always denote a commutative ring.

The following characterization of biregular group rings by W.D. Burgess will be used frequently.

Result 1: [Brg, Proposition] Let A_x denote the stalk of the Pierce sheaf of central idempotents of A at x in $X(R)$. The group ring AG is biregular iff (a) A is a biregular ring; (b) G is a locally normal group with the order of every element in G a unit in A; and (c) the coefficients of every central idempotent of $A_x G$ lift to central elements of A.

For the sake of simplicity, a group G shall be said to satisfy condition LNU with respect to A if G satisfies condition (b) in Result 1.

Mihovski has shown [Mi2; Mi1, Corollary 7] that condition (c) is superfluous in case A satisfies a chain condition on principal two-sided ideals. The condition (c) is also superfluous in the case being investigated.

Remark 2: If the A in Result 1 is an Azumaya ring, then condition (c) is satisfied since A satisfies the central epimorphism condition. (This fact was noted by Burgess.)

The first result may be viewed as an extension to Azumaya rings of a result of O. Villamayor [Vi, Theorem 8] characterizing weakly separable group algebras.

Theorem 3: Let A be a ring with center C and let G be a group. AG is a biregular Azumaya group ring if and only if (i) A is a biregular Azumaya ring and (ii) G is locally normal and the order of every element is invertible in A.

The examples of Burgess [Brg, Example] and Burkholder [Bu2, Example 13] show that not every biregular ring is an Azumaya ring.

In particular, we have the following generalization of the classical Maschke's Theorem:

Corollary 4: Let F be a field and let G be a locally normal group. If the order of every element G is invertible in F, then FG is an Azumaya ring. If char $(F) = 0$, FG is an Azumaya ring for every locally normal group G.

The theorem is proved by a series to reductions.

Lemma 5: Let A and B be rings with center R. If $A \otimes_R B$ is an Azumaya ring and A is an Azumaya ring, then B is an Azumaya ring.

Proof: By [Az, Proposition 2.8], A is a faithfully flat R-module. This allows us to apply [Ra, Proposition 1.11] to obtain that B is a central ideal R-algebra. Let Q be a two-sided ideal of B and let π_* denote the restriction of the natural map $\pi : B \to B/Q$ to the homomorphism (not necessarily onto): $\pi_{Cen(B)} : Cen(B) \to Cen(B/Q)$. Since $A \otimes_R B$ is an Azumaya ring, we have that $1_A \otimes \pi$ is an epimorphism; by [Bo, I.3.1 Proposition 4], π_* is an epimorphism and so B is an Azumaya ring.

Lemma 6: An Azumaya ring A with perfect center C is projective as a C-module and C is a C-direct summand of A.

This lemma follows from [Az, Proposition 2.8 and Corollary 2.14] when one recalls that for a module over a perfect ring the weak and projective dimensions coincide.

The following lemma may be viewed as an extension of [DJ, Lemma 2.2] to Azumaya rings.

Lemma 7: Let A be a ring with center C. The group ring AG is an Azumaya ring if and only if A is an Azumaya ring and CG is an Azumaya ring.

Proof: Suppose that AG is an Azumaya ring. Then A is also an Azumaya ring by [Az, Proposition 2.4].

Now, $AG \simeq A \otimes_C CG$ for any ring A with center C and any group G. If $z(CG)$ denotes the center of CG, we have the following isomorphisms:

$$AG \simeq A \otimes_C CG \simeq (A \otimes_C z(CG)) \otimes_{z(CG)} CG .$$

So by Lemma 5, CG is an Azumaya ring. The converse is readily verified.

Proposition 8: Let R be a commutative ring and G a locally normal group such that the order of each element of G is a unit in R, then every non-central element x of RG can be written as an element of the center plus an element in the two-sided ideal generated by $< rx - xr : r \in RG >$.

Proof: Let g be an arbitrary non-central element of G. Since G is locally normal, the number ν of distinct conjugates of g divides the order of any finite normal subgroup H containing g and so is a unit in G. Let $h_1, \ldots , h_\nu$ be the distinct conjugates of g.

$$\sum_{i=1}^{\nu} (h_i^{-1} g h_i)$$

is a central element of RG. For every g in G, the following equation holds in RG:

$$g = \frac{1}{\nu} \sum_{i=1}^{\nu} [h_i^{-1}(h_i g - g h_i)] + \frac{1}{\nu} \sum_{i=1}^{\nu} (h_i^{-1} g h_i) \ .$$

By linearity, any element x in RG can be written as an element of the center plus the sum of elements of the two-sided ideal.

Applying [Bu1, Lemma 1.1] and [Az, Prop 2.3] in conjunction with this lemma, we have that, whenever G satisfies property LNU with respect to R, RG satisfies the central epimorphism property.

Corollary 9: For a commutative ring R and a locally normal group G with the property that the order of each element g in G is a unit in R, then the center of every homomorphic image of RG is the image of the center of RG.

We are now ready to proceed with the proof of the theorem.

Proof of Theorem 3:

Assume first that AG is a biregular Azumaya ring. Then A is a biregular Azumaya ring since A is a homomorphic image of AG [Az, Proposition 2.4]. The second part (ii) is proved by Burgess [Brg, Proposition].

Conversely, that AG is biregular follows from Result 1 and Remark 2. Since A is biregular, so is its center C; thus CG is biregular by Result 1. Hence CG satisfied the ideal lifting condition since every two-sided ideal is generated by a central idempotent. By Corollary 9, CG satisfies the central epimorphism condition. So CG is an Azumaya ring. Finally, Lemma 7 guarantees that AG is an Azumaya ring.

In particular we have the following corollary which will be used later in considering Azumaya group rings with locally perfect centers.

Corollary 10: Let R be a commutative von Neumann regular ring and G a group with property LNU with respect to R. (i) RG is an Azumaya ring. (ii) If A is any Azumaya ring with center R, then AG is an Azumaya ring.

For an arbitrary commutative ring, the following extension of Maschke's Theorem can be obtained. (It is conjectured that the condition on the conjugacy classes is superfluous.)

Proposition 11: Let R be a commutative ring and G a locally normal group with the property that the order of every element g in G is a unit in R. Assume further that every element g in G is contained in a finite normal subgroup H of G such that the conjugacy class in H of each element h in H is identical to the conjugacy class of h in G. Then RG is an Azumaya ring.

Proof: The central epimorphic condition holds due to Corollary 9. To see that the ideal lifting condition holds, we consider any two-sided ideal I of RG. For an arbitrary element x in I, the number of elements of G in the support of x is finite and so these elements are contained in a finite normal subgroup H of G. By assumption, H may be chosen to satisfy the additional condition that the conjugacy class of each element of H is identical with its conjugacy class in G. Since RH is an Azumaya algebra by Maschke's theorem, $x \in I \cap RH$ is generated by central elements of RH. But $z(RH) = RH \cap z(RG)$ [see Pa, Lemma 4.1.1, p. 113] since the conjugacy classes are equal. Hence x is generated by central elements of RG. RG is, therefore, an Azumaya ring.

Combining Lemma 7 with Proposition 11, we have the more general result:

Corollary 12: Let A be an Azumaya ring. For any group G satisfying property LNU with respect to A and the conjugacy condition of Proposition 11, AG is an Azumaya ring.

Numerous results have been obtained by D. Burkholder [Bu1, Bu2] concerning Azumaya rings whose centers are locally perfect rings. More specialized results may be obtained concerning these rings.

Proposition 13: Let R be a locally perfect commutative ring and G a group satisfying property LNU with respect to A. Then RG is an Azumaya ring with locally perfect center.

Proof: Let $J(R)$ denote the Jacobson radical of R. By [Bu1, Remark 3.5a], $R/J(R)$ is von Neumann regular and so $[R/J(R)]G$ is a biregular Azumaya ring by Theorem 3. Now $J(R)G$ is the kernel of the natural map $\pi : RG \to [R/J(R)]G$. Since the image is biregular, $J(R)G = J(RG)$.

Suppose M is a non-zero left RG-module. Since the RG-module M is naturally an R-module, $J(R)G \cdot M = J(R) \cdot M \neq M$ since $J(R)$ is T-nilpotent [AF, Lemma 28.3, p. 314]. By the same result $J(RG)$ is T-nilpotent. Therefore, since RG satisfies the central epimorphism condition by Corollary 9, the center of RG also has T-nilpotent radical and $z(RG)/J(z(RG)) \simeq z(RG/J(R)G)$ is von Neumann regular. Thus $z(RG)$ is locally perfect.

To see that RG is an Azumaya ring, let M be a maximal two-sided ideal of RG. Then $RG/M \simeq \{[R/J(R)]G\}/\{M/J(R)G\}$ is a simple ring since it is a homomorphic image of the biregular ring $[R/J(R)]G$. Since $z(RG)$ is locally perfect, we may apply Theorem 1.2 of [Bu1] to obtain that RG is an Azumaya ring.

Corollary 14: Let A be an Azumaya ring with locally perfect center and let G be a group satisfying property LNU with respect to A. Then AG is an Azumaya ring with locally perfect center.

The following examples provide concrete examples of Azumaya rings from other known Azumaya rings.

Example 15: Let Z_p denote the localization of the rational integers at the prime p. Let R denote the ring Z_p/p^n for some positive integer n. Let D_k denote the dihedral group of order 2k. Let G denote the direct sum of countably many copies of D_k. Then RG is an Azumaya ring for all $n \geq 1$ and all odd primes p for which p and k are relatively prime. Moreover, RG is biregular iff $n = 1$, p is odd and $(p, k) = 1$. For $n > 1$, $z(RG)$ is locally perfect. If A is an Azumaya algebra such as $M_n(R)$–the full matrix ring over R, then AG is an Azumaya ring when the requirements on n, p and k are met.

Example 16: Let D be any division ring of characteristic 0 and $A_n(D)$ the nth Weyl algebra over D for some positive integer n. By [MR, Theorem 1.3.8(ii)], $A_n(D)$ is a simple ring and is, therefore, an Azumaya ring with center C which is a field. For any locally normal group G, $A_n(D)G$ is an Azumaya ring.

F.R. Demeyer and G.J. Janusz completely characterized those algebras A and groups G for which AG is an Azumaya algebra in [DJ, Theorem I]. No similar result is yet available for Azumaya rings.

Finally, we note that combining Result 1 of Burgess with [DJ, Theorem I] provides the following characterization of biregular Azumaya algebras for contrast with Theorem 3.

Proposition 17: Let A be a ring with center C and G a group. AG is a biregular Azumaya algebra iff (i) A is a biregular Azumaya algebra, (ii) G is locally normal (iii) $[G : Z(G)] < \infty$ and hence G' is finite, and (iv) the order of every normal subgroup is a unit in A.

In Example 15, if $n = 1$, RG is a biregular Azumaya ring but never an Azumaya algebra for $k \geq 4$ since condition (iii) of Proposition 17 fails. For $k = 2$ or 3, D_k is its own center and so AG is an Azumaya algebra.

Acknowledgement: This author wishes to thank Ohio University for the use of their facilities during a portion of the preparation of this paper. In particular, Professors S.K. Jain and S. López-Permouth and another visitor Professor P.F. Smith deserve special thanks for their kind assistance. The author also thanks the referee for pointing out an error in an earlier proof of Proposition 13 and for strengthening Lemma 7.

BIBLIOGRAPHY

AF F. Anderson and K.R. Fuller, *Rings and Categories of Modules*, Springer-Verlag, Berlin-Heidelberg-New York, 1974.

Az G. Azumaya, Separable rings, J. Algebra **63** (1980) 1-14.

Bo N. Bourbaki, Algebre Commutative, Chapitre I, Actualites Sci. Ind. No. 1290, Hermann, Paris, 1962.

Brg W.D. Burgess, A characterization of biregular group rings, Canad. Math. Bull. **21** (1978) 119-120.

Bu1 D.G. Burkholder, Azumaya rings with locally perfect centers, J. Algebra **103** (1986) 606-618.

Bu2 D.G. Burkholder, Azumaya rings, Pierce stalks and central ideal algebras, Comm. Algebra **17** (1989) 103-113.

Bu3 D.G. Burkholder, Products of Azumaya rings and Kochen's map, Comm. Algebra **17** (1989) 115-134.

DI F.R. Demeyer and E.C. Ingraham, Separable Algebras over Commutative Rings, Lecture Notes in Mathematics #181, Springer-Verlag, Berlin-Heidelberg-New York, 1970.

DJ F.R. Demeyer and G.J. Janusz, Group rings which are Azumaya algebras, Trans. Amer. Math. Soc. **279** (1983) 389-395.

Mi1 S.V. Mihovski, Biregular crossed products, J. Algebra **114** (1988) 58-67.

Mi2 S.V. Mihovski, Errata: Biregular crossed products, J. Algebra **117** (1988) 525.

MR J.C. McConnell and J.C. Robson, *Noncommutative Noetherian Rings*, Wiley, London, 1988.

Pa D.S. Passman, *The Algebraic Structure of Group Rings*, Wiley, New York, 1977.

Ra M.L. Ranga Rao, Azumaya, semisimple and ideal algebras, Bull. Amer. Math. Soc. **78** (1972) 588-592.

Vi O.E. Villamayor, On weak dimension of algebras, Pacific J. Math **9** (1959) 941-951.

Computer Sciences Corporation
Integrated Systems Division
304 W. Route 38
Moorestown, NJ 08057

UNIFORM MODULES OVER SERIAL RINGS II

Bruno J. Müller and Surjeet Singh

The purpose of this paper is to reformulate the main result of [5] in an equivalent, but much more natural manner (Theorem 2). We also give explicit characterizations of the relevant ideal $\cap T^n$ (Theorems 6 and 7).

1. BACKGROUND

We shall use [5] freely; this paper is available at present in preprint form. We restate now some of its notations and results:

$N \subset M$ means that N is a proper submodule of M. The letter e is reserved for an indecomposable idempotent of the ring R. A local element is a nonzero element x of some R-module such that xR is a local module, $i.e.$, has a unique maximal submodule. Over a semiperfect ring, and in particular over a serial ring, any element of any module is a sum of local elements. The following are results from [5]:

(1.5) If xR is a local module over a semiperfect ring R, then there is an indecomposable idempotent e such that $xe = x$.

(1.6) (special case). For any ideal I of a right serial ring, and local element x, $x(\cap I^n) = \cap xI^n$ holds. $\cap I^n$ is an almost nonsingular ideal (cf. (2.4) below).

(1.7) Let f be a homomorphism defined on a uniserial module M, and assume $U \subset V \subseteq M$ and $fV \neq 0$. Then $fU \subset fV$ and $fM/fV \cong M/V$.

(1.8) Any proper factor module of a uniserial module is singular.

(2.2) A nonzero projective module over a serial ring is not singular.

(2.4) For an ideal I of a serial ring R, the following are equivalent: (ii) for every indecomposable idempotent e, $eI = 0$ or eR/eI is a nonsingular R/I-module; (iv) for every uniserial module M and $x \in M - MI$, $xI = MI$ holds. (Such ideals I are called almost nonsingular.)

(2.5) If I is an almost nonsingular ideal in a serial ring R, and M a uniserial R-module with $MI \neq 0$, then M/MI is a nonsingular R/I-module.

(3.1) Any two incomparable prime ideals of a right serial ring are co-maximal.

(3.2) Let P, Q be prime ideals of a right serial ring. Assume there is a uniserial module M such that $MP \subset M$ and $MQ \subset M$. Then P, Q are comparable.

(3.4) Every right serial semiprime ring is a direct sum of prime rings.

(3.5) Let Q, P be two incomparable prime ideals of a right serial ring R. If Q is a successor of P, then $ePQ \subset eP \subset eR$ holds for all indecomposable idempotents $e \notin P$.

(3.6) In a serial ring, Q is a successor of P iff $PQ \subset P \cap Q$ or $Q \subset P$.

(3.8) Let Q and P be two incomparable prime ideals of a serial ring.

(i) Q is a successor of P iff P is a left successor of Q;

(ii) If so, then Q and P contain the same prime ideals properly;

(iii) Moreover, Q and P are Goldie and determine each other uniquely.

(4.2) Let S be a Goldie semiprime ideal of a serial ring R, and let M be a uniform S-nonsingular R-module. Then $\cup_{n \in \mathbb{N}} ann_M (S^n)$ is uniserial.

The main result of [5] is

(4.3) Let I be an almost nonsingular ideal of a serial ring R, and let M be a uniform I-nonsingular R-module. Then $ann_M (\cap_{n \in \mathbb{N}} I^n)$ is uniserial.

2. I-NONSINGULAR MODULES

A module M is called I-nonsingular if $ann_M I$ is a nonsingular R/I-module. This property is automatically satisfied if $ann_M I = 0$. In the other case, $ann_M I \neq 0$, we shall completely analyze it in the situation which is of interest here.

We say that a uniform module M "has an assassinator, P" if M has a "P-prime" submodule, i.e., a nonzero submodule N such that $P = ann_R N'$ for all $0 \neq N' \subseteq N$. Such P is automatically prime.

Lemma 1. Let M be a uniform right-module over a right-serial ring R. The following are equivalent:

(i) M is I-nonsingular, with $ann_M I \neq 0$, for some ideal I of R;

(ii) M has an assassinator, P, which is right-Goldie, and $ann_M P$ is a nonsingular ($= torsionfree$)R/P-module.

In this situation, the ideals I with (i) are precisely the ideals $I \subseteq P$ with $eI = eP$ for all $e \notin P$. The only prime ideal with (i) is $I = P$.

In analogy with [4] et al., we shall call a module M as in Lemma 1 $(P-)tame$. (It is not necessary to mention P explicitly since it is uniquely determined.)

Proof of Lemma 1. (ii) is a special case of (i); take $I = P$. Conversely we assume (i). Let $N = Ann_M I$, and consider any A between I and $ann_R N$.

Claim. For any local element $x \in N$, with $x = xe$, one has $ann_R x = eA = eI$. N is a nonsingular R/A-module.

Proof of Claim: As $xI = xA = 0$, we have $eR \supset ann_{eR} x \supseteq eA \supseteq eI$. If $ann_{eR} x \supset eI$, then $xR \cong eR/ann_{eR} x$ is a singular R/I-module, by (1.8), a contradiction. Hence $ann_{eR} x = eA = eI$. We deduce that $xR \cong eR/eA$ is a projective R/A-module, hence not singular (2.2). As x was arbitrary, we conclude that N is a nonsingular R/A-module.

Now we consider an arbitrary $e \notin A$. Then $Ne \neq 0$. Hence there is a local element $x \in N$ with $x = xe$. As before we have $xR \cong eR/eA$. Since $xR \subseteq N$ is R/A-nonsingular, so is eR/eA. We conclude that R/A is a right nonsingular ring.

We quote some statements from [2]: according to (6.13) the prime radical $\sqrt{A}$ of A is a semiprime right-Goldie ideal; by (1.14) R/A is a right-Goldie ring; and hence by (1.35) $\sqrt{A}/A$ is nilpotent.

We conclude that $N' = ann_M A$ is nonzero. By (3.4) there is a (unique) prime ideal P, minimal over $\sqrt{A}$, such that $N'P = 0$. P is also right-Goldie. By the claim (applied to N' instead of N, and with P instead of A), N' is a nonsingular $(= torsionfree)R/P$-module. This means that M is P-nonsingular.

N' is a prime module (since any ideal strictly larger than P is essential modulo P). Therefore, $P = ass\ M$. This establishes (ii).

Next, consider again an arbitrary ideal I with (i). As $P = ass\ M$ is unique maximal among annihilators of nonzero submodules of M, we have $I \subseteq P$. For any $e \in P$, there is local $x = xe \in N'$; then $eI = eP$ by the claim.

Conversely, for an arbitrary ideal $I \subseteq P$ with $eI = eP$ for all $e \notin P$, consider any $0 \neq x \in N'$. Then $0 \neq xe \in N'$ for some $e \notin P$, hence xeR $\cong eR/eP = eR/eI$ is R/I-projective, hence $xe \notin Z_{R/I}(N')$, hence $x \notin Z_{R/I}(N')$. We conclude that N' is a nonsingular R/I-module, $i.e.$, M is I-nonsingular.

Finally, if I is prime, $eP = eI \subseteq I$ implies $e \in I$ or $P \subseteq I$, for all $e \notin P$. At least one $e_0 \notin P$ exists, and $e_0 \in I$ implies the contradiction $e_0 \in P$. We conclude $P \subseteq I$, hence $P = I$.

3. LINKS

Again in analogy with [4] et al., we define *links* between prime ideals of a serial ring R, as follows: $P \leadsto> Q$ means that P and Q are equal or incomparable, and that Q is a successor of P. According to (3.7) and (3.8), this is equivalent to P and Q being equal or incomparable, and $PQ \subset P \cap Q$.

If $P \leadsto> Q$, then (by (3.8) and an easy additional consideration if $P = Q$)P and Q determine each other uniquely. Moreover they are both Goldie if $P \neq Q$. As sets of incomparable primes are finite, cf. (3.3), *cliques* (= link connectivity components) are finite.

It follows that there are only two types of cliques, *linear* ones, $P_1 \leadsto> ... \leadsto> P_t$, and circular ones, $P_1 \leadsto> ... \leadsto> P_t \leadsto> P_1$. We shall reserve the letter T for the semiprime ideal which is the intersection of the clique of the prime ideal P.

Remark. Cliques, as defined here for serial rings, are much better behaved than those in Noetherian rings: they are always finite, links are few, non-trivial cliques are few, and cliques are (by definition) incomparable. However, it may be of interest that cliques need not be Krull homogeneous (if the serial ring R has Krull dimension):

For example, start with a commutative discrete rank two valuation domain A, with maximal ideal m and another nonzero prime ideal p. The localization A_p at p is naturally of rank one, with maximal ideal p. The matrix ring $R = \begin{pmatrix} A & A_p \\ p & A_p \end{pmatrix}$ is serial and prime, with the three nonzero prime ideals $P_1 = \begin{pmatrix} m & A_p \\ p & A_p \end{pmatrix}$, $P_2 = \begin{pmatrix} p & A_p \\ p & A_p \end{pmatrix}$

and $P_3 = \begin{pmatrix} A & A, \\ p & p \end{pmatrix}$. P_2 is linked to P_3, both ways (so the clique is circular). One has $Kd(R) = 2$, $Kd(R/P_2) = 1$ but $Kd(R/P_3) = 0$.

4. THE MAIN RESULT

We prove now our reformulation of (4.3), and demonstrate afterwards that it indeed generalizes (4.3).

Theorem 2. Let M be a uniform P-tame module over a serial ring, and let T be the intersection of the clique of P. Then $ann_M \left(\cap_{n \in \mathbb{N}} T^n \right)$ is uniserial.

Proof. T is Goldie, since P is Goldie, and by (3.8).

(3.4) shows that, for an arbitrary semiprime ideal S, a uniform R/S-module is S-nonsingular iff it is P-nonsingular for some prime ideal P minimal over S. In particular, since M is P-nonsingular, it is T-nonsingular.

Thus (4.2) applies, with $S = T$, and shows that $\cup_{n \in \mathbb{N}} ann_M (T^n)$ is uniserial.

Now consider $0 \neq x \in ann_M (\cap T^n)$. There is $0 \neq xr \in ann_M P \subseteq ann_M T$, hence $xrT = 0$. By (1.6), $xrR \supset 0 = x(\cap T^n) = \cap xT^n$. We obtain t with $xrR \supseteq xT^t$, hence $0 = xrT = xT^{t+1}$. We conclude $\cup ann_M T^n = ann_M (\cap T^n)$.

The following observation shows that Theorem 2 generalizes (4.3):

Lemma 3. Under the assumptions of (4.3), $ann_M \left(\cap_{n \in \mathbb{N}} I^n \right) \subseteq ann_M \left(\cap_{n \in \mathbb{N}} T^n \right)$ if M is P-tame, and $ann_M \left(\cap_{n \in \mathbb{N}} I^n \right) = 0$ if M is not P-tame.

Proof. Case 1: $ann_M I \neq 0$. In the proof of (4.3) one considers ideals $I \subseteq A \subseteq B \subseteq R$ such that $A/I = Z(R/I)$ and $B = \sqrt{A}$. One shows that B is Goldie, M is B-nonsingular, hence P-nonsingular for some minimal prime P of B; in other words that M is P-tame.

Moreover, one establishes $ann_M (\cap I^n) = \cup ann_M (B^n)$. Thus (4.2) applies, with $S = B$. The proof of (4.2) shows more than is stated, namely: let $M_n = ann_M (B^n)$. As long as $0 = M_0 \subset M_1 \subset \cdots \subset M_n$ is strict, there are linked primes $P_1 <\sim \cdots <\sim P_n$, all minimal over B, such that $M_i P_i = M_{i-1} (i = 1, \cdots, n)$.

Therefore, all P_i belong to the clique of $P_1 = P$. Consequently, $P_i \supseteq T$, and $M_n T^n \subseteq M_n P_n \cdots P_1 = 0$. We conclude that $ann_M (B^n) \subseteq ann_M (T^n)$ is valid for all n, and deduce $ann_M (\cap I^n) = \cup ann_M (B^n) \subseteq ann_M (\cap T^n)$.

Case 2: $ann_M I = 0$. In this case, the proof of (4.3) establishes that M is $\cap I^n$-nonsingular. If $ann_M (\cap I^n) = 0$, we are done. If $ann_M (\cap I^n) \neq 0$, by (1.6), Case 1 applies, with I replaced by $\cap I^n$, and shows $ann_M (\cap I^n) \subseteq ann_M (\cap (\cap I^n)^m) \subseteq ann_M (\cap T^m)$.

5. CHARACTERIZATION OF THE IDEAL $\cap T^n$

In view of Theorem 2, it is of great interest to describe $\cap_{n \in \mathbb{N}} T^n$ in terms of P.

We note first that the prime ideals which are properly contained in P, if any, form a chain and are contained in $\cap T^n$.

Indeed, they are comparable by (3.1) and are contained in T by (3.8). If, for a prime $Q \subset P$ and some n, $Q \not\subseteq T^n$, then $eQ \supset eT^n$ for some e, hence $e \in Q$ or $T \subseteq Q$. $e \in Q$ yields $e \in T$, hence $e \in T^m$, hence $eR = eT^m \subset eQ$, a contradiction. $T \subseteq Q$ yields $P_i \subseteq Q \subset P$ for some P_i, a contradiction again. We conclude $Q \subseteq \cap T^n$, as claimed.

Lemma 4. Let R be a serial ring whose prime radical, N, is nilpotent. Then the set of minimal primes is a union of full cliques. The number of these cliques equals the number of summands of a ring-direct decomposition of R into indecomposables.

Proof. Let $P_1, \cdots, P_s, Q_1, \cdots, Q_t$ be the minimal primes of R, where precisely $P_1, \cdots, P_s$ belong to the clique of P_1. Let $A = P_1 \cdots P_s$ and $B = Q_1 \cdots Q_t$. (3.1) implies $A^n + B^n = R$ for every n. As there is no link between P_i and Q_j, they commute by (3.6); hence $A^n B^n = B^n A^n$. Consequently $A^n \cap B^n = A^n B^n$. As $AB \subseteq N$ and $N^n = 0$ for some n, we obtain the ring-direct decomposition $R = A^n \oplus B^n$.

Suppose $P \neq P_1, \cdots, P_s$ is linked to P_i. Then $P \supseteq Q_j \supseteq B^n$, for some j. On the other hand, $P_i \supseteq A^n$. Thus P and P_i commute, in contradiction to (3.6). Therefore, $P_1, \cdots, P_s$ is a full clique.

If R is indecomposable, then $A^n = 0$ and $B = R$, hence no Q_j exists, and there is only one clique. If R is decomposable, then there are obviously no links between minimal primes belonging to different summands and, therefore, each indecomposable summand contributes precisely one clique.

We shall make use of the following result, which was proved by Chatters ([1], Theorem 3.3) under the additional assumption that the ring has Krull dimension. But his proof, verbatim, gives our slightly more general version:

Proposition 5. Let R be an indecomposable serial ring whose prime radical is Goldie and nilpotent. Suppose R has a non-nilpotent ideal X such that $\cap_{n \in \mathbb{N}} X^n = 0$. Then R is prime.

Theorem 6. Let R be an indecomposable serial ring, P a Goldie prime ideal of R, and T the intersection of the clique of P. If the clique is linear, $P_1 \sim> \cdots \sim> P_t$, then $\cap_{n \in \mathbb{N}} T^n = T^t = P_1 \cdots P_t$ is idempotent. If the clique is circular, then either T is nilpotent *or* $\cap_{n \in \mathbb{N}} T^n$ is a Goldie prime ideal.

Remarks. In the linear case, $\cap T^n$ need not be zero; *e.g.*, consider a commutative valuation ring R with nonzero idempotent maximal ideal.

In the circular case, if T is nilpotent, P is a minimal prime, and $\cap T^n = 0$. If $\cap T^n$ is Goldie prime, P is not minimal, and $\cap T^n$ is the largest prime ideal properly contained in P. In particular, then, such a largest prime exists and is Goldie.

Proof. We fix a decomposition $1 = \sum e_i$ into indecomposable orthogonal idempotents.

Case 1: The clique is linear. By (3.6), all P_i commute with each other, except for $P_i P_{i+1} \subset P_i \cap P_{i+1} = P_{i+1} P_i$; moreover, $P_i^2 = P_i$. Therefore, $T^n \supseteq (P_1 \cdots P_t)^n \supseteq P_1^n \cdots P_t^n = P_1 \cdots P_t \supseteq T^t$ for all n. Consequently, $T^n = P_1 \cdots P_t = T^t$ for all $n \geq t$, and the claim follows.

Case 2: For each $e_i \notin T$ there are m_i, n_i such that $e_i T^{m_i} = 0 = T^{n_i} e_i$. Write $\epsilon = \sum \{e_i : e_i \in T\}$; then $1 \neq \epsilon \in T$. For $n = \max\{m_i, n_i\}$, $(1 - \epsilon) T^n = 0 = T^n (1 - \epsilon)$. Therefore, $T^n = \epsilon T^n = T^n \epsilon = \epsilon R = R\epsilon$. Consequently, ϵ is central, hence $\epsilon = 0$, hence $T^n = 0$.

Case 3: The clique is circular, $P_1 \sim> \cdots \sim> P_t \sim> P_1$, and (wlog) there is $e = e_i \notin T$ such that $eT^n \neq 0$ for all n.

We assume, in addition, $eT^n \supset eT^{n+1}$ for some n (this is certainly true for $n = 0$).

By (3.4) there is a unique P_i which annihilates eT^n / eT^{n+1}. Since, by (2.5), eT^n / eT^{n+1} is a nonsingular R/P_i-module, $P_i = \mathrm{ann}_R (eT^n / eT^{n+1})$.

Consider any $x \in eT^n - eT^{n+1}$. By (2.4) we have $xT = eT^{n+1}$. By (1.5) there is e' with $xe' = x$. The epimorphism $f : e'R \to xR$ satisfies $f(e'T^m) = xT^m = eT^{n+m}$ for all $m \geq 1$. Therefore, $\ker f \subseteq \cap e'T^m = e'(\cap T^m)$. Thus $e'R/e'(\cap T^m) \cong xR/x(\cap T^m) = xR/e(\cap T^m)$, using $x(\cap T^m) = \cap xT^m = \cap eT^{n+m} = e(\cap T^m)$ from (1.6). Consequently, $xR/e(\cap T^m)$ is a projective, and hence by (2.2) not singular, $R/\cap T^m$-module.

Moreover, we have $e' \notin P_i$ (since $e' \in P_i$ implies the contradiction $x = xe' \in eT^n P_i = eT^{n+1}$). By (3.5) we know $e'P_i \supset e'P_i P_{i+1} \supseteq e'T^2$. Hence f yields, with (1.7), the strict inclusion $eT^{n+1} = xT = xP_i \supset xP_i P_{i+1} \supseteq xT^2 = eT^{n+2}$.

As above, we obtain some $P_j = \mathrm{ann}_R (eT^{n+1} / eT^{n+2})$. Therefore, $eT^{n+1} P_j \subseteq eT^{n+2} \subset eT^{n+1}$ and $eT^{n+1} P_{i+1} \subset eT^{n+1}$, which implies $P_j = P_{i+1}$ by (3.2).

All of this proves, by induction over n, that $eT^n \supset eT^{n+1}$ holds for all n, that the primes P_i reoccur periodically as the annihilators of the eT^n / eT^{n+1}, and that $eR/e(\cap T^n)$ is nonsingular as $R/\cap T^n$-module.

We show that $e'T^n \neq 0$ holds for all $e' \notin T$. $e' \notin T$ implies $e' \notin P_i$ for some i. There is n such that $P_i = \mathrm{ann}_R (eT^n / eT^{n+1})$. There is $y \in eT^n - eT^{n+1}$ such that $x := ye' \in eT^n - eT^{n+1}$. For the corresponding $f : e'R \to xR$ we have $f(e'T^m) = eT^{n+m} \neq 0$ for all $m \geq 1$. Thus $e'T^m \neq 0$ for all m.

Consequently, all $e' \notin T$ have the same properties as the original e. In particular, $e'R/e'(\cap T^n)$ is a nonsingular $R/\cap T^n$-module. Therefore, $R/\cap T^n$ is a (right) nonsingular ring. By the results ([2], 6.13, 1.14, 1.35), used already in the proof of Lemma 1, we conclude that the prime radical of $R/\cap T^n$ is Goldie and nilpotent.

$T/\cap T^n$ is non-nilpotent, as $eT^n \supset eT^{n+1} \supseteq e(\cap T^m)$.

Finally R/T^2 is indecomposable by Lemma 4, since its minimal primes are precisely the P_i/T^2, and they form *one* clique since all links are preserved due to $T^2 \subseteq P_i P_j$. We deduce that $R/\cap T^n$ is indecomposable, as $\cap T^n$ and T^2 contain the same idempotents.

Altogether, we have now verified all the assumptions of Proposition 5 for $R/\cap T^n$. We conclude that $\cap T^n$ is prime (and Goldie since $R/\cap T^n$ is nonsingular).

6. RINGS WITH KRULL DIMENSION

Theorem 6 can be improved if R has Krull dimension. We recall that a serial ring R has Krull dimension, α, iff α is minimal with $J(\alpha)$ nilpotent (where $J(\alpha)$ is defined inductively by $J(\alpha) = \cap_{\beta < \alpha, \, n \in \mathbb{N}} J(\beta)^n$). In any ring with Krull dimension, any prime ideal is Goldie, the prime radical is nilpotent, and any uniform module has an assassinator (cf. [3], 3.4, 5.1, 8.3).

If A is any ideal, in any ring R with Jacobson radical J, and if $\epsilon \in A + J$ for some idempotent ϵ, then $\epsilon \in A$ [write $\epsilon = a+j$, $r = (1-j)^{-1}$; then $\epsilon = \epsilon(1-j)r = \epsilon r - \epsilon rj = \epsilon(\epsilon - j)r = \epsilon ar \in A$]. Therefore, if R is semiperfect, and if we pick $\epsilon = \epsilon^2 \in R$ such that $\bar{A} = \bar{\epsilon}\bar{R}$ in the semisimple ring $\bar{R} = R/J$, then $\epsilon \in A$; moreover, $1 - \epsilon \in B$ whenever B is another ideal with $A + B = R$.

Theorem 7. Let R be an indecomposable serial ring with Krull dimension, P a prime ideal, and T the intersection of the clique of P. If P is minimal, then T is nilpotent (and hence $\cap T^n = 0$). If P is not minimal, then $\cap T^n$ is the largest prime ideal properly contained in P.

Proof. Since everything follows from Theorem 6 if the clique of P is circular, we assume that it is linear, $P_1 \sim > \cdots \sim > P_t$. Then we know that $A := T^t = P_1 \cdots P_t$ is idempotent. The preceding observation yields an idempotent $1 \neq \epsilon \in A$ such that $1 - \epsilon \in B$ whenever $A + B = R$. For an arbitrary ordinal γ, the following implications hold:

(1) if all P_j contain $J(\gamma)$ and one P_i is minimal over $J(\gamma)$, then all P_j are minimal over $J(\gamma)$;

(2) if all P_j are minimal over $J(\gamma)$, then $J(\gamma + 1) \supseteq A(1 - \epsilon)$;

(3) if all P_j contain $J(\gamma)$ and $J(\gamma) \supseteq A(1 - \epsilon)$, then all P_j are minimal over $J(\gamma)$.

Indeed, for (1): all P_j contain $J(\gamma)^2$, and P_i is minimal over $J(\gamma)^2$. The links are preserved in $R/J(\gamma)^2$ since $J(\gamma)^2 \subseteq P_j P_k$. Hence the claim follows from Lemma 4.

(2) Let B be the product of the minimal primes over $J(\gamma)$ different from P_1, $\cdots$, P_t. Then $AB = BA$ and $A + B = R$, hence $1 - \epsilon \in B$. Consequently, $J(\gamma) \supseteq (AB)^n = AB^n \supseteq A(1 - \epsilon)$ for some n. Therefore, $J(\gamma)^m \supseteq A(1 - \epsilon)$ for all m, hence $J(\gamma + 1) \supseteq A(1 - \epsilon)$.

(3) Pick Q minimal over $J(\gamma)$ such that $1 - \epsilon \in Q$; this is possible since $1 - \epsilon \neq 0$, hence $1 - \epsilon \notin J(\gamma)$. Then $Q \supseteq J(\gamma) \supseteq A(1 - \epsilon)$ implies $Q \supseteq A$, hence $Q \supseteq P_i \supseteq J(\gamma)$ for some i. Therefore, $Q = P_i$ is minimal over $J(\gamma)$. By (1) all P_j are minimal over $J(\gamma)$.

Next let $\beta_j = Kd(R/P_j)$ and $\beta = \beta_i = \max \beta_j$. We prove, by induction for all $\gamma \geq \beta$, (i) $J(\gamma) \supseteq A(1 - \epsilon)$ and (ii) all P_j are minimal over $J(\gamma)$.

Indeed: we have $J(R/P_j)(\beta_j) = 0$, and this implies easily $J(\beta) \subseteq J(\beta_j) \subseteq P_j$. By [3], 7.2 P_i is minimal over $J(\beta)$. Then (1) and (2) establish (ii) and (i) for $\gamma = \beta$.

If (i) and (ii) hold for all $\beta \leq \delta < \gamma$, then (i) for γ follows from (2) if γ is a non-limit ordinal, and is trivial if γ is a limit ordinal. Then (ii) follows from (i) and (3).

Let $Kd(R) = \alpha$. Then $J(\alpha + 1) = 0$, hence $A(1 - \epsilon) = 0$. As $\epsilon \in A$, we obtain $A = A\epsilon = R\epsilon$. By symmetry we also have $A = \epsilon R$. Thus ϵ is central, hence $\epsilon = 0$, hence $A = 0$.

REFERENCES

[1] A.W. Chatters, Serial rings with Krull dimension, Glasgow Math. J. **32** (1990), 71-78.

[2] A.W. Chatters and C.R. Hajarnavis, *Rings with Chain Conditions*, Pitman, 1980.

[3] R. Gordon and J.C. Robson, *Krull Dimension*, Amer. Math. Soc. Memoirs **133** (1973).

[4] A.V. Jategonkar, *Localization in Noetherian rings*, London Math. Soc. Lecture Note Series **98**, Cambridge University Press, 1986.

[5] B.J. Müller and Surjeet Singh, Uniform modules over serial rings, McMaster Math. Reports, Preprint Series No. **6** (1989/90), 25.

McMaster University
Hamilton, Ontario L8S 4K1
CANADA

and

Department of Mathematics
Kuwait University
P.O. Box 5969
13060 Safat, KUWAIT

LINKS BETWEEN PRIME IDEALS OF A SERIAL RING
WITH KRULL DIMENSION

Mary H. Wright*

For the basic definitions of a serial ring and the Krull dimension of a module, denoted $K.\dim M$, the reader is referred to any of [1, 6 or 7] and to [2]. In [8], the prime ideals of a serial ring with Krull dimension were characterized in terms of transfinite powers of the Jacobson radical. In this article, we give a definition of a right link $Q \sim > P$ where Q and P are prime ideals of a serial ring with Krull dimension. Our definition reduces to that of Jategaonkar in the case of a Noetherian serial ring (see [3]). We shall show how links between prime ideals are connected with successors of cliques of local projective modules discussed in [6]. It turns out that links between prime ideals of a serial ring with Krull dimension share some nice properties with certain classes of Noetherian rings, especially with regard to such questions as the second layer condition and the incomparability condition. In spite of very nice properties of the link graph of a serial ring with Krull dimension, some desirable aspects of the theory over Noetherian rings do not extend to non-Noetherian serial rings.

It is assumed throughout that R denotes a serial ring with Krull dimension. Modules are unitary right modules unless specified otherwise. The symbol $\subset$ denotes strict inclusion. The Jacobson radical of a ring S is denoted by $J(S)$, or just J if the ring S is clear from the context. The injective hull of a module is denoted by $E(M)$. Recall from [8] that the transfinite powers, $I(\alpha)$, of an ideal I in R are defined by: $I(0) = I$ and for $\alpha > 0$, $I(\alpha) = \cap\{I(\beta)^n | \beta < \alpha$ and $n \in \mathbf{N}\}$. Clearly, any serial ring is semiperfect; we shall assume that $1 = \Sigma\{e | e \in F\}$ is a fixed decomposition of 1 as a sum of local orthogonal idempotents. Our first result follows easily from [8, Prop. 3.1 and 3.2].

PROPOSITION 1. If P is a prime ideal of R, $\beta = K.\dim. R/P$, $U = \{e \in F | e \notin P\}$ and $u = \Sigma\{e | e \in U\}$, then $P = uJ(\beta) \oplus (1-u)R = J(\beta)u \oplus R(1-u)$. Furthermore, $\forall \gamma < \beta$, $\forall n \in \mathbf{N}$, $eP = eJ(\beta) \subset eJ(\gamma)^n$ and $Pe = J(\beta)e \subset J(\gamma)^n e$. Conversely, if $e \in F$ and β is an ordinal such that $\forall \gamma < \beta$, $\forall n \in \mathbf{N}$, $eJ(\beta) \subset eJ(\gamma)^n$, then there exists a prime ideal P such that $eP = eJ(\beta)$; also K.dim. $R/P = \beta$. In fact, P is uniquely determined by such e and β.

Given prime ideals Q and P, we say that there is a *right link from Q to P via A*, denoted $Q \sim >_A P$, provided A is an ideal such that $QP \subseteq A \subset Q \cap P$, $(Q \cap P/A)_R$ embeds in a finite direct sum of copies of $E(R/P)$ and $_R(Q \cap P/A)$ embeds in a finite direct sum of copies of $_R E(R/Q)$. Obviously a right link from Q to P via A is identical with a left link from P to Q via A.

We shall give a characterization of a (right) link between a pair of prime ideals which brings to light the close connection between this notion and that of an α-successor of a β-clique of local projective modules. It will be useful to recall certain facts concerning α-cliques. For each ordinal $\alpha \geq 0$, $\{eR | e \in F\}$ is partitioned into α-*cliques*. For $\alpha = 0$, these are simply the singletons $\{eR\}$, $e \in F$. For $\alpha > 0$, the α-cliques are defined to be the equivalence classes in the equivalence relation generated by $\mathcal{R}$, where

$(eR, fR) \in \mathcal{R}$ if $\exists \gamma < \alpha$ such that eR is a projective cover of some (cyclic) submodule of $fJ(\gamma)/fJ(\gamma)^2$. The unique α-clique containing eR will be denoted by $C(\alpha, eR)$. Obviously, if $\beta \le \alpha$, then $C(\beta, eR) \subseteq C(\alpha, eR)$.

An α-clique A is called an α-successor of a β-clique B (and B is called a β-predecessor of A) provided there exists $eR \in A$ and $fR \in B$ such that $fJ(\beta)J(\alpha) \subset f(\beta) \cap fJ(\alpha)$ and eR is a projective cover of some (cyclic) submodule of $fJ(\beta)/fJ(\beta)J(\alpha)$. If this is the case and $\alpha > \beta$, then $fJ(\alpha) = fJ(\beta)$; otherwise, by [8, Lemma 1.1], $fJ(\beta)J(\alpha) = fJ(\alpha)$. Hence, for any α and β, if a given β-clique B is a β-successor of an α-clique A, then $\forall fR \in B$, $fJ(\beta) \subseteq fJ(\alpha)$.

For $\alpha > 0$, a given α-clique A may be either circular or linear [7]. If α is a nonlimit ordinal, A is a disjoint union of $(\alpha - 1)$-cliques $A_1, \ldots, A_k$, which may be indexed so that for $i = 1, \ldots, k-1$, A_{i+1} is the $(\alpha - 1)$-successor of A_i. If the $(\alpha - 1)$-successor of A_k is A_1, we call A a *circular* α-clique. If not, then A_k has no $(\alpha - 1)$-successor and A_1 has no $(\alpha - 1)$-predecessor. In this case, A is called a *linear* α-clique. When α is a limit ordinal, there is the smallest ordinal γ, necessarily a nonlimit ordinal, such that A is already a γ-clique. Either $(\forall \delta \in [\gamma, \alpha), \exists \epsilon \in [\delta, \alpha), A$ is circular as an ϵ-clique) or $(\exists \delta \in [\gamma, \alpha), \forall \epsilon \in [\delta, \alpha), A$ is linear as an ϵ-clique). In the first case, A is called a *circular* α-clique (and necessarily $\forall \delta \in [\gamma, \alpha), \exists \epsilon \in [\delta, \alpha), A$ is its own ϵ-successor). In the second case, A is called a *linear* α-clique.

If for some $U \subseteq F$, $A = \{eR | e \in U\}$, denote by A^* the set $\{Re | e \in U\}$. The following summarizes some important results concerning α-cliques. Parts (a), (b), (d), (e) and (f) are implicit in [6, Prop. 3]. Part (c) is [7, Prop. 1.1 and Prop. 1.2]. Part (g) is an easy consequence of [8, Prop. 3.1].

PROPOSITION 2. Let R be a serial ring with Krull dimension.

(a) Given an α-clique A and an ordinal β, A has at most one β-successor (resp. β-predecessor). Furthermore, if A is linear, its α-successor, if it exists, must be circular.

(b) If A is an α-clique of right R-modules, then A^* is an α-clique of left R-modules. Furthermore, a β-clique B is a β-successor (resp. β-predecessor) of A iff B^* is a β-predecessor (resp. β-successor) of A^* in $R - Mod$.

(c) If $e \in F$ is a local idempotent and α an ordinal such that $eJ(\alpha) \ne 0$, then $C(\alpha, eR)$ is circular iff $\forall \gamma < \alpha$, $\forall n \in \mathbf{N}$, $eJ(\alpha) \subset eJ(\gamma)^n$.

(d) If $\beta < \alpha$ and $e \in F$ is such that $eJ(\alpha)J(\beta) \subset eJ(\alpha)$, then the α-clique generated by the projective covers of cyclic submodules of $eJ(\alpha)/eJ(\alpha)J(\beta)$ is linear.

(e) If $\alpha \ge \beta$ and $e \in F$, if $0 \subset eJ(\alpha)J(\beta) \subset eJ(\alpha)$, then the set of projective covers of cyclic submodules of $eJ(\alpha)/eJ(\alpha)J(\beta)$ is a β-clique. If this β-clique is circular, the hypothesis that $eJ(\alpha)J(\beta) \ne 0$ may be dropped.

(f) For $e, e' \in F$, if $C(\alpha, eR) = C(\alpha, e'R)$ is circular, and if both $eJ(\alpha)$ and $e'J(\alpha)$ are nonzero, then eR is a projective cover of some submodule of $e'R/e'J(\alpha)$.

(g) Let $e \in F$, $P \in \operatorname{Spec} R$, $\delta = K.\dim. R/P$. Then $C(\delta, eR) = \{fR | f \in F - P\}$.

We mention that two incomparable prime ideals, P and Q, of a serial ring are comaximal; indeed, for each $e \in F$, either $eP \subseteq eQ \subseteq Q$ (from which follows $e \in Q$) or $eQ \subseteq eP \subseteq P$ (from which follows $e \in P$).

PROPOSITION 3. Let Q and P be prime ideals of R; let $\gamma = K.\dim. R/Q$ and $\delta = K.\dim. R/P$. The following are equivalent:

(a) $Q \sim>_A P$ for some ideal A containing QP.

(b) There exists local idempotents $f, g \in F$ and an element $x = fxg \in Q \cap P - QP$ such that

   (i) $K.\dim. Rx/Qx = \gamma$ and $K.\dim. xR/xP = \delta$ and

   (ii) $fQg = fJ(\gamma)g \supset fJ(\gamma)J(\delta)g = fQPg \subset fJ(\delta)g = fPg$.

(c) $QP \subset Q \cap P$ and $\forall f, g \in F$, if $fQPg \subset f(Q \cap P)g$, then fQ/fQP is P-torsion free in $Mod - R$ and Pg/QPg is Q-torsion free in $R - Mod$.

Proof: Assume that (a) holds. Certainly there exists $x = fxg$ in $Q \cap P - QP$ for some $g, f \in F$. For any such x, there are commutative diagrams in $R - Mod$ and $Mod - R$:

$$Qf = J(\gamma)f \subset Rf$$
$$\downarrow \qquad \downarrow$$
$$QPg = Qx = J(\gamma)x \subset Rx \subseteq Pg = J(\delta)g \subset Rg$$

and

$$gP = gJ(\delta) \subset gR$$
$$\downarrow \qquad \downarrow$$
$$fQP = xP = \quad xJ(\delta) \subset xR \subseteq fQ = fJ(\gamma) \subset fR.$$

Indeed, $(xR + A)/A_R$ is P-torsion free and $_R(Rx + A)/A$ is Q-torsion free. There is a canonical homomorphism $gR/gP \longrightarrow\!\!\!\!\!\rightarrow xR/xP \cong xR/xR \cap A$. Since $xR/xR \cap A$ is P-torsion free, it follows that $gR/gP \cong xR/xP$. Similarly, $Rf/Qf \cong Rx/Qx$. Clearly, $xP \subseteq fQP$. Proper inclusion would imply that $xR/xR \cap A$ is P-torsion, a contradiction. Assertion (b)(ii) follows trivially.

Assume that (b) holds. Let f' be an idempotent in F such that $f'QP \subset f'(Q \cap P)$. Let $x' = f'x' \in Q \cap P - QP$. Since $x' \notin QP$, we have $f' \notin Q$. By Prop. 2, $f'R$ is a projective cover of some submodule of fR/fQ; hence fQ/fQP is a homomorphic image of $f'Q/f'QP$. From this it follows easily that $f'Q/f'QP$ is P-torsion free as well as essential in $f'R/f'QP$. Thus $Q \cap P/QP$ embeds in a finite direct sum of copies of $E(R/P)$. By a symmetric argument, $_R(Q \cap P/QP)$ embeds in a finite direct sum of copies of $_R E(R/Q)$. Hence (b) $\Rightarrow$ (a) and (c). Trivially, (c) $\Rightarrow$ (b).

The commutative diagrams which arose in the proof of Prop. 3 evidently imply that if f and g satisfy (b), then $C(\delta, gR)$ is the δ-successor of $C(\gamma, fR)$. The next result makes the connection between linked primes and successors of cliques more precise.

PROPOSITION 4. Suppose e and f are local idempotents (in F), α and β are ordinals such that $A = C(\alpha, eR)$ and $B = C(\beta, fR)$ are circular, B is the successor of A and $eJ(\alpha)/eJ(\alpha)J(\beta)$ is nonsingular as an $R/J(\beta)$-module. Then there are prime ideals Q and P such that $eJ(\alpha) = eQ$, $fJ(\beta) = fP$ and $Q \sim> P$.

Proof: By Prop. 2, we may assume without loss of generality that $\alpha \geq \beta$. (If not, pass to $R - Mod$). There is an element $x = exf \in eJ(\alpha) - eJ(\alpha)J(\beta)$ and a commutative diagram:

$$fJ(\beta) \subset fR$$

$$\downarrow \qquad \downarrow$$

$$xJ(\beta) = eJ(\alpha)J(\beta) \subset xR \subseteq eJ(\alpha) \subset eR$$

By Prop. 1, there exist prime ideals Q and P such that $eJ(\alpha) = eQ$ and $fJ(\beta) = fP$. The set of projective covers of cyclic submodules of $eJ(\alpha)/eJ(\alpha)J(\beta)$ is precisely $C(\beta, fR)$ (Prop. 2(e)). It follows easily that $xJ(\beta) = xP = eJ(\alpha)P = eQP$, that eQPf $\subset eJ(\alpha)f \subseteq eJ(\beta)f$, and that K.dim. $xR/xP = \beta$.

If K.dim. $Rx/Qx < \alpha$, then $J(\epsilon)^m x \subseteq Qx \subseteq QP$ for some $\epsilon < \alpha$ and $m \in \mathbf{N}$. Now $eJ(\epsilon)^m \supset eQ$ and $C(\alpha, eR)$ is circular; hence $eJ(\epsilon)^m \supset yR \supset eQ$ for some $y = eye \in eR$. Furthermore, $Ann(yR/eQP) = Ann(eR/eQP)$. But then $x \in Ann(eR/eQP)$; in particular, $ex = x \in eQP$, contradiction. This shows that $Q \sim> P$.

PROPOSITION 5. In a serial ring with Krull dimension, if Q, P and P' are prime ideals such that $Q \sim> P$ and $Q \sim> P'$ (resp. $P \sim> Q$ and $P' \sim> Q$), then $P = P'$.

Proof: Let $\gamma = K.\dim. R/Q$, $\delta = K.\dim. R/P \leq \delta' = K.\dim. R/P'$. There are local idempotents $f, f', g, g' \in F$ and elements $x, x' \in fR, f'R$, respectively, such that the following diagrams commute:

$$Qf \subset Rf \qquad\qquad Qf' \subset Rf'$$
$$\downarrow \quad \downarrow \qquad\qquad \downarrow \quad \downarrow$$
$$QPg \subset Rx \subseteq Pg \subset Rg \qquad QP'g' \subset Rx' \subseteq P'g' \subset Rg'$$

$$gP \subset gR \qquad\qquad g'P' \subset g'R$$
$$\downarrow \quad \downarrow \qquad\qquad \downarrow \quad \downarrow$$
$$fQP \subset xR \subseteq fQ \subset fR \qquad f'QP' \subset x'R \subseteq f'Q \subset f'R$$

Now $C(\gamma, fR) = C(\gamma, f'R)$. By uniqueness of δ'-successors, $C(\delta', gR) = C(\delta', g'R)$. Using projective covers and appealing to Prop. 2, we may assume that $g = g'$ and $f = f'$. It follows immediately that P and P' are comparable and (since $\delta \leq \delta'$) $P' \subseteq P$. If $\delta = \delta'$, obviously $P = P'$.

If $\gamma \leq \delta < \delta'$, and if $fJ(\delta') \subset fJ(\gamma)$, then we get $fJ(\gamma)J(\delta')g = fJ(\delta')g$, which contradicts Prop. 3. Hence, $fJ(\gamma) = fJ(\delta) = fJ(\delta')$. This implies $fJ(\gamma) \subseteq fJ(\delta') \subseteq fJ(\delta)^2 \subseteq fJ(\gamma)J(\delta)$, again a contradiction.

Similarly, if $\delta < \delta' \leq \gamma$, then $J(\delta)g = J(\delta')g = J(\gamma)g \Rightarrow J(\gamma)J(\delta)g = J(\gamma)g$, a contradiction.

If $\delta < \gamma < \delta'$, then $fJ(\gamma) = fJ(\delta')$; hence $C(\delta', fR)$ is linear. Also $fJ(\delta') \supset fJ(\delta')J(\delta)$. This implies that the δ'-successor of $C(\delta', fR)$ is linear, which is impossible.

PROPOSITION 6. Let P be a non-minimal prime ideal of a serial ring with Krull dimension. Then there exist prime ideals $P = P_1, P_2, \ldots, P_n$ such that $P_1 \leadsto> P_2 \leadsto> \ldots \leadsto> P_n \leadsto> P_1$. Furthermore, the primes P_i are incomparable for $i = 1, \ldots, n$.

Proof: Since P is non-minimal, there exists a prime ideal Q and a local idempotent $e_1 \in F$ such that $e_1 Q \subset e_1 P \subset e_1 R$. Let $\alpha = K.\dim. R/Q$, $\gamma_1 = K.\dim. R/P$, $P_1 = P$. We proceed inductively on $\alpha = K.\dim. R/Q$. If $\alpha = 1$, the result is clear.

Assuming $\alpha > 1$, let $\gamma_2 \leq \alpha$ be the smallest ordinal such that $e_1 J(\gamma_1)J(\gamma_2) \subset e_1 J(\gamma_1)$. If $\gamma_2 = \alpha$, then (inductively) $e_1 J(\gamma_1) = e_1 J(\beta)^m \forall \gamma_1 \leq \beta < \alpha, \forall m \in \mathbf{N}$. In this case, $e_1 J(\gamma_1) = e_1 J(\alpha)$, contradiction. Hence, $\gamma_2 < \alpha$.

If $\gamma_1 < \gamma_2$ and $e_1 J(\gamma_1) \supset e_1 J(\gamma_2)$, then $\forall \delta < \gamma_2, \forall n \in \mathbf{N}, e_1 J(\gamma_2) \subset e_1 J(\gamma_1) = e_1 J(\gamma_1)J(\delta)^n \subseteq e_1 J(\delta)^n$. Then there exists a prime ideal Q' such that $e_1 J(\gamma_2) = e_1 Q' \subset e_1 P \subset e_1 R$. It follows that $Q' \subset P$ with $K.\dim. R/Q' < \alpha$. By the induction hypothesis and Prop. 5, we are done in this case.

It remains to consider the cases $\gamma_1 \geq \gamma_2$ or $\gamma_1 < \gamma_2$ with $e_1 J(\gamma_1) = e_1 J(\gamma_2)$. In either case, the set of projective covers of cyclic submodules of $e_1 J(\gamma_1)/e_1 J(\gamma_1)J(\gamma_2)$ is a γ_2-clique (Prop. 2), which is circular by minimality of γ_2.

We may select $e_2 \in F$ and $y = ye_2 \in e_1 J(\gamma_1) - e_1 J(\gamma_1)J(\gamma_2)$. Since $y \notin Q$, $e_2 \notin Q$. There exists a prime ideal P_2 such that $e_2 J(\gamma_2) = e_2 P_2 \supset e_2 Q$ and there is a commutative diagram

$$e_2 Q = e_2 J(\alpha) \subset e_2 P_2 = e_2 J(\gamma_2) \quad \subset \quad e_2 R$$

$$\Big\downarrow \qquad\qquad\qquad \Big\downarrow \qquad\qquad\qquad \Big\downarrow$$

$$e_1 Q = e_1 J(\alpha) \subset yJ(\gamma_2) = e_1 J(\gamma_1)J(\gamma_2) \subset yR \subseteq e_1 J(\gamma_1) \subset e_1 R.$$

By Prop. 4, $P_1 \leadsto> P_2$.

This procedure may be iterated, producing a sequence of linked prime ideals $P_1 \leadsto> P_2 \leadsto> \ldots$. where (by induction and Prop. 5) all P_i are minimal among prime ideals properly containing Q. Associated with each P_i is the subset $T_i = F - P_i \subseteq F$. Since F is finite, there is some $f \in F$ and some pair $i \neq j$ such that $f \in T_i \cap T_j$. It then follows that $P_i = P_j$. In view of Prop. 5, the proof is now complete.

Following the standard terminology for Noetherian rings, we shall call the connected components of the link graph of Spec R *cliques* (of prime ideals). It will always be clear from the context whether we mean a clique of prime ideals or an α-clique of local projective modules. In any case, Props. 3 and 4 show that the two notions are closely related.

COROLLARY 7. Cliques of prime ideals in a serial ring with Krull dimension are incomparable.

Proof: If not, we have prime ideals Q and P belonging to the same clique, such that $Q \subset P$. But then Prop. 6 yields incomparable primes $P_1 = P, P_2, \ldots, P_n$ such that the clique of P is $\{P_1, \ldots, P_n\}$, which cannot contain Q.

Example 1. While all non-minimal prime ideals of a serial ring with Krull dimension belong to a finite, cyclic clique, this need not be the case for a minimal prime ideal. Let T denote the ring of integers localized at a prime ideal (p), M the unique maximal ideal of T and U the injective hull of T/M (in $Mod - T$). Let $B = End(U)$. Let $S = \left(\begin{smallmatrix} B & U \\ 0 & T \end{smallmatrix}\right)$ with the usual matrix addition and multiplication. Then S is a serial ring with Krull dimension 1, $J = J(S) = \left(\begin{smallmatrix} J(B) & U \\ 0 & M \end{smallmatrix}\right) = K \cap L$, where $K = \left(\begin{smallmatrix} B & U \\ 0 & M \end{smallmatrix}\right)$, $L = \left(\begin{smallmatrix} J(B) & U \\ 0 & T \end{smallmatrix}\right)$; $J(1) = \left(\begin{smallmatrix} 0 & U \\ 0 & 0 \end{smallmatrix}\right) = P \cap Q$, where $P = \left(\begin{smallmatrix} B & U \\ 0 & 0 \end{smallmatrix}\right)$, $Q = \left(\begin{smallmatrix} 0 & U \\ 0 & T \end{smallmatrix}\right)$. K, L, P and Q are the only prime ideals of S; P and Q are minimal. Let $f = \left(\begin{smallmatrix} 1 & 0 \\ 0 & 0 \end{smallmatrix}\right)$, $g = \left(\begin{smallmatrix} 0 & 0 \\ 0 & 1 \end{smallmatrix}\right)$. Since $fQ = fQJ$ and $Q = Q^2$, the only chance for Q to belong to a circular clique is if $Q \rightsquigarrow P \rightsquigarrow Q$. But fQ is Artinian; hence P-torsion in this case. Thus the clique of Q is $\{Q\}$, an isolated point in the link graph.

Example 2. One respect in which links in a serial ring with Krull dimension do not behave as nicely as in some Noetherian rings concerns the Krull dimension of the factor rings corresponding to linked primes. Indeed, it is possible to have primes $Q \rightsquigarrow P$ with K.dim. $R/Q \neq$ K.dim. R/P. To see this, let T be as in Example 1 and let K denote the classical, simple Artinian quotient ring of T. Let $S = \left(\begin{smallmatrix} T & K \\ 0 & K \end{smallmatrix}\right)$, $f = \left(\begin{smallmatrix} 1 & 0 \\ 0 & 0 \end{smallmatrix}\right)$, $g = \left(\begin{smallmatrix} 0 & 0 \\ 0 & 1 \end{smallmatrix}\right)$. Again, S is a serial ring with Krull dimension one. We have $J = J(S) = \left(\begin{smallmatrix} M & K \\ 0 & 0 \end{smallmatrix}\right)$ and $J(1) = \left(\begin{smallmatrix} 0 & K \\ 0 & 0 \end{smallmatrix}\right) = Q \cap P$, where $Q = \left(\begin{smallmatrix} 0 & K \\ 0 & K \end{smallmatrix}\right)$ and $P = \left(\begin{smallmatrix} T & K \\ 0 & 0 \end{smallmatrix}\right)$. Then $Q \rightsquigarrow P$. But K.dim. $R/Q \neq$ K.dim. R/P.

Following [3], we call a subset X of Spec R *stable* if $P \in X$ and $(Q \rightsquigarrow P$ or $P \rightsquigarrow Q)$ implies $Q \in X$. It is well known that over a Noetherian ring A, a finite, stable, incomparable subset $X \subseteq$ Spec A gives rise to classical localization provided that X satisfies one additional condition, namely the so-called "second layer condition". Recall that the *assassinator* of a uniform module M over a ring with Krull dimension is the unique maximal element in the set of prime ideals that annihilates some nonzero submodule of $M[2]$. Over any ring with Krull dimension, every uniform module M has an assassinator, denoted by ass M. Modifying [3] very slightly for a serial ring with Krull dimension, we say that a subset X in

Spec R *satisfies the right second layer condition* if for each $P \in X$, if Q is the assassinator of some uniform submodule $U \subseteq E(R/P)/Ann$ P, then U_R is Q-torsion free. We say that X satisfies the second layer condition if it satisfies both the right and the left second layer conditions.

PROPOSITION 8. Any set of prime ideals in a serial ring with Krull dimension satisfies the second layer condition.

Proof: Let P be any prime ideal, $E = E(R/P)$, $A = Ann_E P$. Let U be any uniform submodule of E/A and let $Q = ass\ U$. Without loss of generality, we may assume $Q = Ann\ U$ and $U = mR + A/A$ where $m = mf$ for some $f \in F$. Then $f \notin Q$. Hence $fQ = fJ(\gamma)$, where $Q = K.\dim.\ R/Q$, and $\forall \delta < \gamma$, $\forall \in \mathbb{N}$, $fJ(\gamma) \subset fJ(\delta)^n$. Since $Q = ass\ U$, it follows that $mQ = mR \cap A \subset mJ(\delta)^n \forall \delta < \gamma$, $\forall n \in \mathbb{N}$.

The following is a corollary to the proof. The reader should compare with [3, Lemma 6.1.6].

COROLLARY 9. For any prime ideal P of R, let $E = E(R/P)$ and $A = Ann_E P$. Then a prime Q is the assassinator of some uniform submodule of E/A iff $Q \sim> P$.

Proof: ($\Rightarrow$) With the same notation as in the proof of Prop. 8, we have, for any $y \in mQ$, a commutative diagram:

$$
\begin{array}{ccc}
fR & \twoheadrightarrow & mR \\
\cup & & \cup \\
fJ(\gamma) = fQ & \twoheadrightarrow & mR \cap A = mQ \\
\cup & & \cup \\
gR \quad \longrightarrow \quad xR & \twoheadrightarrow & yR \\
\cup \qquad\qquad \cup & & \cup \\
gJ(\delta) = gP \twoheadrightarrow fQP & \twoheadrightarrow & 0
\end{array}
$$

Clearly, $fJ(\gamma)J(\delta) \subset fJ(\gamma)$. If $\gamma \geq \delta$, we certainly have $fJ(\gamma)J(\delta) \subset fJ(\delta)$. If $\gamma < \delta$ and $fJ(\gamma)J(\delta) = fJ(\delta)$, then fR and gR belong to the same δ-clique. Since this δ-clique is circular, $f \notin P$. But then $fP = fJ(\delta) = fJ(\gamma)J(\delta) \subseteq fQP \Rightarrow mP = 0$, a contradiction. Thus we have shown that the circular δ-clique $C(\delta, gR)$ is the δ-successor of the circular γ-clique $C(\gamma, fR)$. By Prop. 4, $Q \sim> P$.

($\Leftarrow$) By Prop. 3, if $Q \sim> P$, there is a commutative diagram (for some $f, g \in F$)

$$
\begin{array}{c}
gP \subset gR \\
\downarrow \qquad \downarrow \\
xP = fQP \subset xR \subseteq fQ \subset fR
\end{array}
$$

where xR/xP is P-torsion free and fR/fQ is Q-torsion free. Without loss of generality, we may assume $QP = 0$. We wish to show that Q is the assassinator of some uniform submodule of $E(R/P)/Ann\ P$. It suffices to check that $fQ = Ann_{fR} P$. As in the proof of Prop. 4, for any $r \in fR - fQ$, there exists $s = fsf$ such that $rR \supset sR \supset fQ$ and $fR \cong sR$; hence $rP \neq 0$.

The upshot of Prop. 6, Cor. 7 and Prop. 8 is that all cliques of prime ideals in a serial ring with Krull dimension are finite and incomparable, and satisfy the second layer condition. The finiteness guarantees that any clique $X \subseteq \text{Spec } R$ also satisfies the *intersection condition*: $\cap \{C(P) | P \in X\} = C(\cap \{P/P \in x\})$, where for any ideal $I \subseteq R$, $C(I)$ denotes the set of elements of R which become regular in R/I. As was shown in [3, Thm. 7.1.5], a stable, incomparable set X of prime ideals in a Noetherian ring A such that X also satisfies the second layer condition and the intersection condition, must give rise to a classical localization. In particular, for each simple A_X-module S, every finitely generated submodule of the A_X-injective hull of S is annihilated by a product of elements of X. The following example shows that this need not happen over a serial ring with Krull dimension.

Example 3. Let T denote the ring of integers localized at a nonzero prime ideal, let $M = J(T)$, let $U = E(T/M)$ and let R be the set of all pairs $\{(a,x) | a \in T, x \in U\}$ with component-wise addition and the following multiplication: $(a,x)(b,y) = (ab, ay + xb)$. Then $J(R) = \{(m,x) | m \in M, x \in U\}$ and $J(1) = \{(0,x) | x \in U\}$. $R \subseteq E(R/J(R))$. Then $X = \{J(R)\}$ is stable and incomparable, and satisfies both the second layer conditon and the intersection condition. In fact, $C(J)$ is trivially left and right localizable, $R_J = R$. But $E(R/J)$ does not have the property that every finitely generated submodule is annihilated by a power of J.

* This research was partially supported by National Science Foundation grant #DMS-8906425.

REFERENCES

[1] A.W. Chatters, Serial rings with Krull dimension, Glasgow Math. J. **32** (1990) 71-78.

[2] R. Gordon and J.C. Robson, Krull dimension, Mem. of the Amer. Math. Soc. #133; Providence, Rhode Island, 1973.

[3] A.V. Jategaonkar, Localization in Noetherian rings, London Math. Soc. Lecture Note Series, #98; Cambridge University Press, Cambridge, 1986.

[4] S. Singh, Serial right Noetherian rings, Can. J. Math. **36** (1984) 22-37.

[5] R.B. Warfield, Jr., Serial rings and finitely presented modules, J. Algebra **37** (1975) 187-222.

[6] M.H. Wright, Krull dimension in serial rings, J. Algebra **124** (1989) 317-328.

[7] ------ , Uniform modules over serial rings with Krull dimension, Commun. in Alg., to appear.

[8] ------ , Prime serial rings with Krull dimension, Commun. in Alg., to appear.

Department of Mathematics
Southern Illinois University at Carbondale
Carbondale, Illinois 62901

SEMIPRIME MODULES AND RINGS

John Dauns

A unital right R-module M is *torsion free* (abbreviation: t.f.) if its singular submodule ZM is $ZM = 0$. Essential submodules are denoted by " $<<$ " and $E(M)$ is the injective hull of M, e.g. $M << E(M)$. A t.f. module D is *discrete* if it contains an essential direct sum of uniform submodules, while C is *continuous* if it contains no uniform submodules. A nonzero torsion free module W is *atomic* if for any $0 \neq y \in W$, there exists an embedding $W \hookrightarrow E(\bigoplus\{yR \mid \Gamma\})$ for some index set Γ. Two atomic modules V and W are equivalent, written as $V \sim W$, if $W \hookrightarrow E(\bigoplus\{V \mid I\})$ and $V \hookrightarrow E(\bigoplus\{W \mid J\})$ for some index sets I and J. Let $\Omega = \{\tau, \rho, \ldots\}$ denote the equivalence classes of torsion free atomic modules. An atomic module is either uniform or continuous, and thus $\Omega = \Omega^C \cup \Omega^D$, $\Omega^C \cap \Omega^D = \emptyset$ where Ω^D consists of equivalence classes of uniform modules, and Ω^C the continuous ones. A continuous compressible module is atomic and hence defines an equivalence class in Ω^C. A t.f. module N is *molecular* if every nonzero submodule of N contains an atomic submodule. Lastly a t.f. module B is *bottomless* if it contains no atomic submodules. Each $\tau \in \Omega$ defines a unique intrinsic submodule $M_\tau = \sum\{W \mid W \leq M, W \in \tau\} \subset M$ which is a complement submodule of M. If $\tau \in \Omega^D$, then M_τ is the complement closure $M_\tau = (\bigoplus_\alpha U_\alpha)^-$ of an essential direct sum $\bigoplus_\alpha U_\alpha << M$ of uniform submodules $U_\alpha \in \tau$. Also, $A, B, D < M$ are right complement submodules.

Theorem I shows that every t.f. module M contains an essential direct sum of three completely unique and algebraically very different submodules $A \oplus B \oplus D << M$, where D is discrete, $A \oplus B$ is continuous, B is bottomless, and $A \oplus D$ is molecular. In particular A is continuous molecular. Moreover $\sum\{M_\tau \mid \tau \in \Omega\} = \bigoplus_\Omega M_\tau << A \oplus D$, and $A \oplus D$ is the complement closure $A \oplus D = (\bigoplus_\Omega M_\tau)^-$. It is surprising that in certain contexts continuous atomic modules behave in the same way as the uniform ones. The same proofs simultaneously cover both kinds of modules. Usually, the proofs from the uniform case do not generalize. For example, it was shown in [D6; p.8, 2.10(iii)] for $A \oplus B \oplus D <<M = R_R$, that $E(D) \cong \Pi\{E(R_\tau) \mid \tau \in \Omega^D\}$ as a right R-module and as a ring. Here this is generalized (5.5 Cor.1 to Theorem I) to any t.f. modules $A \oplus B \oplus D << M$ by showing that $(EA) \oplus (ED) = \Pi\{E(M_\tau) \mid \tau \in \Omega\}$. From this it follows that any t.f. injective continuous molecular module such as $E(A)$ is a full direct product $E(A) = \Pi\{E(M_\tau) \mid \tau \in \Omega^C\}$.

A strongly prime module M whose annihilator $M^\perp$ is not large in M is automatically a t.f. prime atomic module. The strongly prime modules are a good source of examples of continuous t.f. prime atomic modules and particularly rings. It is by now evident from the above that in this study generalized prime and semiprime modules, complement submodules and complement closures of modules, and certain subdirect products play a central role.

The objectives of this article are far more extensive than to merely list a few main results (Theorems I, II, and III). Here we want to outline a general theory of complement submodules, semiprime and prime modules, and certain subdirect product representations of modules so that a large part of the proofs of Theorems I, II, and III would become special cases of a theory that is useful in other contexts, and moreover is of intrinsic interest by itself. The main results, most of which are new, are labeled either as propositions or as theorems.

The author feels that a ring R is understood in some sense if essential direct sums can be identified inside R, and we know how the separate pieces multiply together. For $M = R_R$ torsion free and for the previously defined submodules $A \oplus B \oplus D << R_R$, the continuous part of the ring is defined as $C = (A \oplus B)^- < R$. All of these turn out to be ideals $A, B, C, D, R_\tau \triangleleft R, \tau \in \Omega$. Then Corollary 2 to Theorem I (5.6) shows how they multiply:

$$A \oplus B \oplus D << C \oplus D << R, \ A \oplus B << C; \ CD = DC = 0, \ AB = BA = 0;$$

$$\bigoplus_{\tau \in \Omega} R_\tau << A \oplus D << C; \ R_\tau \oplus R_\tau^\perp << R; \ R_\tau R_\rho = 0, \ \tau \neq \rho \in \Omega.$$

Theorem II shows that a torsion free semiprime ring R induces a suprising amount of primeness and semiprimeness conditions on its complement submodules, including the A, B, C, D, and R_τ listed above. For example, any atomic right ideal $W < R$ is a prime right R-module, and not merely semiprime.

The last Theorem III proves that a necessary and sufficient condition for a ring R to be torsion free, semiprime, and right molecular is that it be an essentially dense subdirect product $R \cong \tilde{R} \subset \Pi\{R_i \mid i \in I\} = T$ of right nonsingular, prime right atomic rings R_i with $\bigoplus(R_i \cap \tilde{R}) << \tilde{R}_R << T_R$. A result of Levy ([L; p.66, Theorem 3.2]) then shows that this subdirect product representation is unique in the strongest possible sense: $I \cong \Omega$, and up to isomorphism, and the R_i are the previous rings $R_i \cong R/R_\tau^\perp, \ \tau \in \Omega$. This theorem is best possible; counterexamples show that the molecular - atomic part of the hypothesis cannot be removed (Counterexamples 7.1). To the best of the author's knowledge this theorem is new even for discrete rings. Examples (7.2) show that there also actually exist continuous rings which satisfy these conditions (for Theorem III).

Some of the motivation behind Theorem III was that the torsion free uniform modules are a special case of a bigger class of modules – the atomic modules. Maybe other theorems which presently are known to hold only for uniform module, have analogues to the bigger class of atomic modules. However, the proofs will not generalize from the uniform to the atomic case in any straightforward way. The class of t.f. atomic prime rings is also strictly bigger than the class of strongly prime rings (see 7.4).

1. FUNDAMENTALS

The following basic classes of modules are defined: torsion free, discrete, continuous, atomic, molecular, bottomless, prime, semiprime, strictly prime and strongly prime. The atomic modules are further distinguished into discrete atomic and continuous atomic ones.

1.1. Notation and Definitions. Here modules M are right unital over a ring R with $1 \in R$. Submodules are denoted by $<$ or $\leqq$, while $<<$ denotes large or essential submodules. If $A < B$ is not large, we write $A <\!\!\!\!\not< B$. For $K < M$ and $m \in M$, $m^\perp = \{r \in R \mid mr = 0\} < R$. Set $m^{-1}K = (m + K)^\perp = \{r \in R \mid mr \in K\} < R$. For a subset $X \subset M$, define $X^\perp = \{r \in R \mid xr = 0 \text{ for all } x \in X\} = \{r \mid Xr = 0\}$. Then $M^\perp = \{r \mid Mr = 0\} \triangleleft R$, where "$\triangleleft$" denotes ideals in R as well as in other rings. Set $\ell(X) = \{r \in R \mid rX = 0\}$.

An ideal $I \triangleleft R$ is an *annihilator ideal* if $I = X^{\perp}$ for some nonempty subset $X \subseteq R$. Always, $I \subseteq \ell(I)^{\perp}$. From $X \subseteq \ell(I)$ we get that $I = X^{\perp} \supseteq \ell(I)^{\perp}$. Thus $I = \ell(I)^{\perp} = \ell(X^{\perp})^{\perp} = X^{\perp}$. An annihilator ideal I is a maximal annihilator ideal if $\{0\}^{\perp} = R$ is the only annihilator ideal which properly contains I. Observe that in a semiprime ring the left and right annihilators of a two sided ideal are equal.

The singular submodule of M is $ZM = Z(M) < M$, where $ZM = \{m \in M \mid m^{\perp} << R\}$. The second singular submodule is located between $ZM \subset Z_2 M \subset M$, and is defined by $Z[M/(ZM)] = (Z_2 M)/(ZM)$. Then $ZM = 0$ iff $Z_2 M = 0$. A module M is *torsion* if $M = Z_2 M$. What is more important for this article is that it is *torsion free* if $ZM = 0$ (abbreviation: t.f.). Here ZR, $Z_2 R \triangleleft R$ denote the right singular submodules of R_R. A submodule $K < M$ is a *complement* submodule of M if there exists some $B < M$ such that K is maximal with respect to $K \cap B = 0$, equivalently, if K has no proper essential extensions inside M. The injective hulls of right R-modules over R are denoted both by "$\widehat{}$" and "E" as $E(M) = EM = \widehat{M}$.

The cardinality of any set I whatever is denoted by $|I|$. The Goldie dimension of any module M - - denoted by $\mathrm{Gd}\, M$ - - is the supremum of all cardinals $|I|$ such that M contains the direct sum of $|I|$ nonzero submodules, i.e.

$$\mathrm{Gd}\, M = \sup\{|I| \mid \exists \bigoplus_{i \in I} V_i \leq M,\ V_i \neq 0\}.$$

If $\mathrm{Gd}\, M$ is not an inaccessible cardinal, then $\mathrm{Gd}\, M$ is attained, that is there exists a direct sum as above with $|I| = \mathrm{Gd}\, M$ maximal. (See [DF].)

The term "type" denotes a special kind of classes of modules and is used to distinguish these from very different kinds of classes of modules used here, such as the prime and semiprime ones.

1.2. Types of Modules. For torsion free modules A, B, C, D, define $A \prec B$ is $A \hookrightarrow E(\bigoplus\{B \mid J\}) = E(\bigoplus_J B)$ for some (in general infinite) index set J. This is a quasi-order and defines an equivalence relation " $\sim$ " where $A \sim C$ if $A \prec C$ and $C \prec B$, i.e. iff $A \hookrightarrow E(\bigoplus_J B)$ and $B \hookrightarrow E(\bigoplus_I A)$ for some index sets I and J. Every equivalence class $[A] = \{C \mid C \sim A\}$ is called a *type*. It has been shown that all the types $\Xi(R) = \{[A], [B], [C], \ldots\}$ form a set ([D9; p.62, 3.15(i)]). The quasi-order induces a partial order on the equivalence classes $\Xi(R)$, where $[A] \leq [B]$ if $A \prec B$, that is if $A' \prec B'$ for some (or equivalently all) $A' \in [A]$, $B' \in [B]$. Note that $0 = [(0)] \leq [A] = [EA]$.

A t.f. module is *continuous* if it contains no uniform submodules, and *discrete* if it contains an essential direct sum of uniform submodules. A nonzero t.f. module W is *atomic* if $[W]$ is an atom in the poset $\Xi(R)$, that is if $0 \neq p \leq [W]$, then $p = [W]$. Thus an atomic module is either continuous or discrete but never both.

More generally, a module N is *molecular* if (i) N is torsion free and (ii) every nonzero submodule of N contains an atomic one. There are two important disjoint subclasses of molecular modules, the discrete molecular and the continuous molecular ones. In general a molecular module N contains an essential direct sum $A \oplus D << N$ where A is continuous and D discrete. (See Theorem I). Although the terms "discrete" and "discrete molecular" by definition mean the same thing, sometimes the latter will be used to distinguish and contrast the discrete molecular modules from the continuous ones. At the other extreme of the module spectrum, a t.f. continuous module B is *bottomless* if B contains no atomic submodules.

The set of atoms $\Omega = \Omega(R) = \{\tau, \rho, \ldots\} \subset \Xi(R)$ is a disjoint union $\Omega = \Omega^C \cup \Omega^D$ where $\Omega^D = \{[U] \mid U_R$ is t.f. uniform$\}$ while Ω^C are the types represented by t.f. continuous atomic modules. The equivalence classes of t.f. continuous compressible modules belong to Ω^C. If U and V are t.f. uniform, then $[U] = [V]$ if and only if $\widehat{U} \cong \widehat{V}$.

1.3. Definition. For any t.f. module M and any element $\alpha \in \Xi(R)$, define a unique intrinsic submodule $M_\alpha \leq M$ by $M_\alpha = \sum\{V \mid V \leq M, [V] \leq \alpha \in \Xi(R)\}$. If there do not exist any $0 \neq V \leq M$ with $[V] \leq \alpha$, then $M_\alpha = 0$.

In particular, if $ZR = 0$, and $M = R_R$, and $\tau \in \Omega$, then $R_\tau = \sum\{U \mid U \leq R, U \in \tau\}$ is the sum of all the atomic right ideals U of type τ.

The usual theory of prime and semiprime rings and right ideals can be done more generally in a module context so that the results for rings are corollaries for the special case $M_R = R_R$. (See [D4] and [D5].)

1.4. Definition. A submodule $K < M$ is *prime* if for any $m \in M$ and $t \in R$

$$mRt \subseteq K \implies \text{either } m \in K \text{ or } Mt \subseteq K.$$

The submodule $K < M$ is *semiprime* if for any $m \in M$ and $s \in R$

$$msRs \subset K \implies ms \in K.$$

The module M itself is said to satisfy any primeness condition, such prime or semiprime, if $(0) < M$ is a prime or semiprime submodule. If $1 \in R$, any $L < R$ is a prime or semiprime right ideal in the usual sense if and only if $L < R$ is a prime or semiprime right R-module.

1.5. Definition. For any $K < M$, define $\Gamma(K) = \{x \in M \mid \forall\ c \in R, xc \in K \implies Mc = 0\}$. Then $K < M$ is *strictly prime* if

$$\forall\ m \in M \setminus K, \ mR \cap \Gamma(K) \neq \emptyset.$$

Thus M is strictly prime if and only if for any $0 \neq m \in M$ there exists an $m^* \in R$ such that $(mm^*)^\perp = M^\perp$.

The next primeness condition has been used in [HL] and [GHL].

1.6. Definition. A submodule $K < M$ is a *strongly prime* if for any $m \in M$, there exists an integer $n = n(m)$ and $s_1, \ldots, s_n \in R$ such that

$$\bigcap_{i=1}^{n} (ms_i)^{-1} K = M^\perp.$$

The set $\{s_1, \ldots, s_n\}$ is called an *insulator* for m. A finite set of elements of R that is an insulator for every $m \in M$ is called a *uniform insulator*.

1.7. Remarks. (1) Strict prime $\implies$ strong prime $\implies$ prime $\implies$ semiprime.

(2) We could have defined a module M to be semiprime iff the intersection of all the prime submodules of M is zero. This was not done because we wanted a definition of

semiprimeness which reduces to the usual definition in case $M_R = R_R$. For an R-module M

$$\bigcap \{V \mid V \leq M \text{ is prime}\} = 0 \implies M \text{ is semiprime.}$$

Whether the converse holds is an open question.

(3) Any intersection of prime (or semiprime) submodules is a prime (or semiprime submodule).

(4) Any direct product of semiprime modules is semiprime. A direct product $\Pi\{W_\gamma \mid \gamma \in \Gamma\}$ of prime modules W_γ, $\gamma \in \Gamma$, is a prime module provided all the W_γ have the same annihilator $W_\gamma^\perp = (\bigoplus W_\gamma)^\perp$. A finite direct sum of strongly prime modules is strongly prime.

2. COMPLEMENT CLOSURES

Complement submodules $C < M$ are easier to work with for at least two reasons. An essential extension $L << M$ remains essential in the quotient $(L + C)/C << M/C$. Secondly, if $ZM \subseteq C$, then also $Z_2 M \subseteq C$ and moreover also $Z(M/C) = 0$. For certain submodules $K \subset M$ the complement closure $K \subset \overline{K} \subset M$ of K is defined. The most frequently used properties of $\overline{K}$ are developed, and its connection with the second singular submodule explained.

2.1. Complement Producing Lemma. For a module M with $ZM = 0$ and any subset $Y \subset M$, $Y^\perp < R$ is a right complement.

Proof. Let $B \leq R$ be maximal with respect to $Y^\perp \oplus B << R$. Then take $A < R$ maximal with respect to $Y^\perp \subseteq A$ and still with $A \cap B = 0$. If $a \in A \setminus Y^\perp$, then $Y^\perp << Y^\perp + aR$. The latter implies that $a^{-1}Y^\perp << R$. But $Ya \neq 0$ whereas $(Ya)(a^{-1}Y^\perp) = 0$. Thus $Ya \subseteq ZM$ is a contradiction. Hence $Y^\perp < R$ is a complement.

Some facts are listed which will allow us to work with complement closures. These facts are not only used later on but are frequently useful in other contexts and hence of independent interest by themselves. For some proofs and more details, see [D9; pp.52-55, 1.1-1.11], [D4; p.165, 3.1-3.9], and [D8; p.2, 1.2]. Moreover, these facts are listed in the order in which they can be most economically and easily proved.

2.2. Definition. For a submodule $ZM \subseteq K < M$, the *complement closure* $\overline{K}$ in M is defined by $Z(M/K) = \overline{K}/K$. The use of the notation "$\overline{K}$" automatically presupposes M and that $ZM \subseteq K$, and thus that $ZM \subseteq K \subseteq \overline{K} \leq M$.

2.3. Observation. For any modules $ZM \subseteq K < M$ and any $x \in M$,

$$x^{-1}K << R \iff K << K + xR.$$

The complement closure of a module $K < M$ is completely determined by the set $M \setminus \overline{K}$, and thus the next lemma gives an elementwise characterization of $\overline{K}$.

2.4. Lemma. For any module M and any submodule $ZM \subseteq K < M$,

$$M \setminus \overline{K} = \{x \in M \mid \exists\, t \in R,\ xtR \neq 0,\ \text{but}\ xtR \cap K = 0\}.$$

Note that in general, complement submodules of a module M are not closed under finite intersections.

2.5. Proposition. Suppose that $ZM \subseteq K < M$ and $Z(M/K) = \overline{K}/K$. Then

(1) $\overline{K} = \{x \in M \mid K << K + xR\} = \{x \in M \mid x^{-1}K << R\}$.

(2) $K << \overline{K}$; $\overline{K}$ is the unique smallest complement submodule of M containing K.

(3) $\overline{K} = \cap\{C \mid K \subseteq C, \ C \leq M \text{ is a complement}\}$; in particular $(\overline{K})^- = \overline{K}$.

(4) $Z(M/\overline{K}) = 0$; $\overline{K} < M$ is the unique smallest submodule of $C \leq M$ containing $K \subseteq C$ such that the quotient module M/C is torsion free.

(5) $K = 0 \iff \overline{K} = 0$.

(6) If $K < M$ is fully invariant, then so is also $\overline{K} < M$.

(7) In particular, if $K \triangleleft R$ is any ideal with $ZR \subseteq K$, then $\overline{K} \triangleleft R$.

(8) If $K = ZM$, then $\overline{K} = Z_2 M$. Hence $Z_2(EM) = E(Z_2 M)$. In particular, $Z_2 R \triangleleft R$.

Frequently it is important to know the behavior of complement submodules under module homomorphisms.

2.6. Lemma. Suppose that $L << A$, $N << B$ are modules and that $f : A \longrightarrow B$ is any homomorphism with $ZB \subseteq fL$. Then the following hold.

(i) $f^{-1}N << A$.

(ii) $fL << fA$.

(iii) If $C \leq B$ is a complement with $fL \subseteq C$, then also $fA \subseteq C$.

(iv) If $ZB = 0$, then $\ker f < A$ is closed and $Z_2 A \subseteq \ker f$.

(v) Suppose that $\hat{f} : EA \longrightarrow B$ is any extension of f and that $D < B$ with $fA \cap D = 0$. Then also $\hat{f}(EA) \cap D = 0$.

If eventually we will take the complement closure of a module, then we may replace the original module with any essential submodule. Therefore in proofs involving complement modules, it is useful to be able to produce essential extensions of modules as below.

2.7. Observation. For a torsion free module M, the following hold:

(i) $L << R$, $K << M \Longrightarrow KL << ML$;

(ii) $L_1 << L_2 < R \Longrightarrow ML_1 << ML_2$;

(iii) $K << M$, $L < R \Longrightarrow KL << ML$.

Proof. (i) If not, $KL \oplus N << M$ for some $0 \neq N \leq M$. Since $K << M$, there exists $0 \neq n \in N \cap K$. Since $n^{-1}K \cap L << R$, and M is t.f., there exists $0 \neq nb \in n[n^{-1}K \cap L]$ with $b \in n^{-1}K \cap L$. But since $b \in L$ and $nb \in K$, we have $nb \in KL \cap N = 0$, a contradiction.

(ii) It suffices to show that for any $0 \neq \xi = \sum\limits_{i=1}^{n} m_i \lambda_i \in ML_2$ with $m_i \in M$ and $\lambda_i \in L_2$, $\xi R \cap ML_1 \neq 0$. From $L_1 << L_2$ it follows that $D = \bigcap\limits_{i=1}^{n} \lambda_i^{-1} L_1 << R$. Since M is t.f., $0 \neq \xi D \subseteq ML_1$. Thus $ML_1 << ML_2$.

(iii) This follows from (ii) with $L = L_1 = L_2$.

3. PRIMENESS AND COMPLEMENTS

The objective of this section is to study various primeness conditions on submodules $K < M$ and particularly what these conditions imply about the complement closure $\overline{K} < M$.

Not all primeness conditions carry over to complement closures. For example, weak semiprimeness does not ([D4; p.161, 1.9 and p.162, 1.13]).

3.1. Proposition. Suppose that $K < M$ is a right R-submodule with $ZM \subseteq K$ and $K << \overline{K}$ its complement closure (2.2). Then the following hold.

(i) $K < M$ is prime $\implies \overline{K} < M$ is prime.

(ii) $K < M$ is semiprime $\implies \overline{K} < M$ is semiprime.

(iii) $K < M$ is strictly prime $\implies \overline{K} < M$ is strictly prime.

(iv) $K < M$ is strongly prime $\implies \overline{K} < M$ is strongly prime.

Proof. We omit the proof of (i) and (ii). (iii) Let $m \in M \setminus \overline{K}$. We need to show that $mR \cap \Gamma(\overline{K}) \neq \emptyset$. In view of $\overline{K} <\!\!\not< \overline{K} + mR$, there exists a nonzero cyclic submodule such that $(ma_0 + k_0)R \oplus \overline{K} \leq M$, for some $a_0 \in R$, $k_0 \in K$. Since $K << M$ is strictly prime, $ma_0 r_1 \in \Gamma(K)$ for some $r_1 \in R$. Suppose that $c \in R$ with $(ma_0 + k_0)r_1 c \in \overline{K}$. We need to show that $Mc \subseteq \overline{K}$. But $(ma_0 + k_0)r_1 c \in (ma_0 + k_0)R \cap \overline{K} = 0$, or $ma_0 r_1 c = -k_0 r_1 c \in K$. From $ma_0 r_1 \in \Gamma(K)$ we conclude that $Mc \subseteq K \subset \overline{K}$. Thus $ma_0 r_1 \in \Gamma(\overline{K})$.

(iv) For $m \in M \setminus \overline{K}$ as before, $\overline{K} \neq (ma_0 + k_0)R \oplus \overline{K} \leq M$. The strong primeness of $K < M$ implies that there exist $s_1, \ldots, s_n \in R$ such that $\bigcap_1^n (ma_0 s_i)^{-1} K = M^\perp$. Now $r \in (ma_0 s_i)^{-1} \overline{K}$ iff $ma_0 s_i r \in \overline{K}$ iff $(ma_0 + k_0)s_i r \in (ma_0 + k_0)R \cap \overline{K} = 0$, or iff $ma_0 s_i r = -k_0 s_i r \in K$. Always $(ma_0 s_i)^{-1} K \subseteq (ma_0 s_i)^{-1} \overline{K}$, and hence $ma_0 s_i^{-1} \overline{K} = ma_0 s_i^{-1} K$. Consequently $\bigcap_1^n (ma_0 s_i)^{-1} \overline{K} = \bigcap_i^n (ma_0 s_i)^{-1} K = M^\perp$ as required.

3.2. Corollary. Let W_γ, $\gamma \in \Gamma$, be a family of torsion free modules. Then the following hold:

(i) $\forall \ \gamma \in \Gamma$, W_γ is semiprime $\implies \prod_{\gamma \in \Gamma} \widehat{W}_\gamma$ is semiprime;

(ii) $\forall \ \alpha, \beta, \gamma \in \Gamma$, all the W_γ are prime with the same annihilator. $W_\alpha^\perp = W_\beta^\perp \implies \prod_{\gamma \in \Gamma} \widehat{W}_\gamma$ is prime.

3.3. Lemma. Let R be a semiprime ring with $ZR = 0$ and let $K < R$. Define $J = \{\xi \in \widehat{R} \mid \widehat{K}\xi = 0\}$. Then

(i) $\widehat{R}$ is a t.f. semiprime ring;

(ii) $K^\perp = \overline{K}^\perp = \widehat{K}^\perp$;

(iii) $J = E(K^\perp)$.

Proof. (i) Suppose that $0 \neq \xi \in \widehat{R}$ with $\xi \widehat{R} \xi = 0$. Since $ZR = 0$, $\xi(\xi^{-1}R) \neq 0$. But then $\xi(\xi^{-1}R)R\xi(\xi^{-1}R) = 0$ contradicts the semiprimeness of R.

(ii) Assume that $K^\perp \neq \widehat{K}^\perp$ for some $K < R$. From $K \subset \overline{K} \subset \widehat{K}$ we get that $\widehat{K}^\perp \subseteq \overline{K}^\perp \subseteq K^\perp$. Then by 2.1, both $K^\perp$ and $\widehat{K}^\perp$ are right complements. Hence $\widehat{K}^\perp \oplus L << K^\perp$ for some $0 \neq L < R$. In a semiprime ring R, for any $K < R$, $K^\perp K = 0$. Thus $LK = 0$. If it were the case that $L\widehat{K} = 0$ in the semiprime ring $\widehat{R}$, then $(\widehat{K}L)\widehat{K}L = 0$, hence $\widehat{K}L = 0$, and $L \subseteq \widehat{K}^{\perp}$, a contradiction. Therefore $L\widehat{K} \neq 0$. Take any $\eta \in \widehat{K}$ with $L\eta \neq 0$. By 1.6 (i) $\eta^{-1}K << R$, and thus $L\eta(\eta^{-1}K) \subseteq LK = 0$ gives the contradiction that $L\eta \subseteq Z(\widehat{R}_R) = 0$. Therefore $\widehat{K}^{\perp} = K^{\perp}$, and hence $K^{\perp} = \overline{K}^{\perp} = \widehat{K}^{\perp}$.

(iii) It follows from the definition of J and $\widehat{K}$ that $J \cap R = \widehat{K}^{\perp}$, and hence $J \cap R = K^{\perp}$ by (ii). From $J \cap R << J$ we deduce that $J = E(J \cap R) = E(K^{\perp})$.

3.4. Corollary. In a t.f. semiprime ring R, if $L << K \leq R$ are right ideals with L large in K, then $L^{\perp} = K^{\perp}$.

3.5. Proposition. For any ring R with $ZR = 0$, the following are all equivalent.

(i) R is semiprime.

(ii) Whenever $A, B \leq R$ are nonzero right ideals such that any cyclic submodule of A is isomorphic to some cyclic submodule of B, then $BA \neq 0$.

(iii) $\forall\ 0 \neq A,\ 0 \neq B \leq R,\ \widehat{A} \cong \widehat{B} \Longrightarrow BA \neq 0$.

Proof. (iii) $\Longrightarrow$ (i). For any $0 \neq x \in R$, take $A = B = xR$. Since $xR(xR) \neq 0$ by (iii), R is semiprime.

(i) $\Longrightarrow$ (ii). If not then $BA = 0$ for some A, B as in (ii). For any $0 \neq a \in A$, by hypothesis $a^{\perp} = b^{\perp}$ for some $b \in B$. Then $a^{\perp} <\!\!\!\not< R$ and $a^{\perp} \oplus C << R$ for some $0 \neq C \leq R$. From $bRaR \subseteq BA = 0$ it follows that also $bCaC = 0$. Thus $CaC \subseteq b^{\perp} \cap C = 0$. Since R is semiprime, $aCaC = 0$, and hence $aC = 0$. Thus $C \subseteq a^{\perp}$ is a contradiction.

(ii) $\Longrightarrow$ (iii). Let $\phi : \widehat{A} \longrightarrow \widehat{B}$ be an isomorphism, and take any $0 \neq a \in A$, and hence $0 \neq \phi a \in \widehat{B}$. It follows from $Z\widehat{B} = 0$ and $(\phi a)^{-1}B << R$ that $0 \neq (\phi a)[(\phi a)^{-1}B] = \phi[a(\phi a)^{-1}B]$. Hence there exists an $x \in a(\phi a)^{-1}B \subseteq A$ with $0 \neq \phi x = y \in B$. Since ϕ is an isomorphism, $xR \cong yR$. Then $0 \neq yRxR \subseteq BA$ by (ii) (with $A \longleftrightarrow xR$, $B \longleftrightarrow yR$).

One of our later objectives will be to deduce that in certain semiprime rings R, many ideals of R are either prime rings, or satisfy the even stronger condition that they are prime right R-modules.

3.6. Observations. Let R be a semiprime (or prime) ring and $0 \neq N \triangleleft R$. Then

(i) N is a semiprime (or prime) right R-module; and

(ii) N is a semiprime (or prime) ring.

The semiprime version of the next proposition will be used later.

3.7. Proposition. Suppose that R is a prime (semiprime) ring and $N \triangleleft R$ is any ideal with $ZR \subseteq N$. Then $\overline{N} \triangleleft R$ is a prime (semiprime) ideal of R.

Proof. By 2.3 (7), $\overline{N} \triangleleft R$. Take any $P < R$ with $\overline{N} \oplus P << R$. If $\overline{N} \triangleleft R$ is not prime, then $xRy \subseteq \overline{N}$ for some $x, y \in R \setminus \overline{N}$. Since $x \notin \overline{N}$, necessarily $x^{-1}N <\!\!\!\not< R$ by 2.3 (1), and hence the first inclusion in $x^{-1}N \subset x^{-1}(N \oplus P) << R$ is proper. For any $c \in [x^{-1}(N \oplus P)] \setminus x^{-1}N$, $0 \neq xc = x_1 + x_2$, where $x_1 \in N$ and $0 \neq x_2 \in P$. Similarly, $0 \neq yd = y_1 + y_2$ with $y_1 \in N$ and $0 \neq y_2 \in P$. Let $r \in R$ be arbitrary. Then the first three terms in $xry = x_1ry_1 + x_1ry_2 + x_2ry_1 + x_2ry_2$ belong to N because $N \triangleleft R$. But $xRy \subseteq \overline{N}$, and hence $x_2ry_2 \in \overline{N} \cap P = 0$. Thus $x_2Ry_2 = 0$ which contradicts the primeness of R. In the semiprime case in the above proof take $x = y$ in which case $c = d$ and $x_2 = y_2$.

3.8. Proposition. Suppose that R is any semiprime ring with $ZR = 0$ and that $N \triangleleft R$ is any ideal whatsoever. Then the following hold:

(i) $N \oplus N^{\perp} << R$; and

(ii) $N^{\perp\perp} = \overline{N}$.

(iii) $\forall \; Q = \overline{Q} \triangleleft R, \; N \oplus Q << R \Longrightarrow Q = N^{\perp}$.

Proof. (i) Let $N \oplus N^{\perp} \oplus D << R$ for $D < R$. Then $DN \subseteq D \cap N = 0$. By semiprimeness, $D \subseteq \ell(N) = N^{\perp}$ and hence $D = 0$. Thus $N \oplus N^{\perp} << R$.

(iii) Clearly, $Q \subseteq N^{\perp}$. Suppose that $Q \subsetneq N^{\perp}$. Since by 2.1 both are right complements, $Q <\!\!\not< N^{\perp}$, and $W \oplus Q << N^{\perp}$ for some $0 \neq W < R$. The fact that $N \oplus Q << R$ and the modular law show that

$$0 \neq (W \oplus Q) \cap [N \oplus Q] = J \oplus Q << W \oplus Q, \quad \text{where}$$
$$0 \neq J = (W \oplus Q) \cap N.$$

By semiprimeness, $N(W + Q) = 0$ implies that also $(W + Q)N = 0$. But then $J^2 = 0$ contradicts the semiprimeness of R.

(ii) By 3.3 (ii), $N^{\perp} = \overline{N}^{\perp}$. Hence (i) becomes $\overline{N}^{\perp} \oplus \overline{N} << R$. By 2.4 (7), $\overline{N} \triangleleft R$. Now use (iii) with N replaced by $\overline{N}^{\perp}$ and with $Q = \overline{N}$ to conclude that $N^{\perp\perp} = \overline{N}^{\perp\perp} = \overline{N}$.

We briefly summarize some of the facts proved about right ideals, ideals, complements, and annihilators in a t.f. semiprime ring.

3.9. Proposition. Suppose that R is a semiprime ring with $ZR = 0$ and that $K < R$ is any right ideal and $N \triangleleft R$ is any ideal. Then the following hold:

(i) $K^{\perp} = \overline{K}^{\perp}$;

(ii) $\forall \; 0 \neq a, \; b \in R, \; a^{\perp} = b^{\perp} \Longrightarrow aRbR \neq 0$;

(iii) $\overline{N} \triangleleft R$ is a semiprime ideal,

(iv) $N^{\perp\perp} = \overline{N}, \; N \oplus N^{\perp} << R$;

(v) $\forall \; P \oplus Q << R, \; P = \overline{P} \triangleleft R, \; Q = \overline{Q} \triangleleft R \Longrightarrow P^{\perp} = Q$.

(vi) $\widehat{R}$ is a t.f. semiprime ring; $K^{\perp} = \widehat{K}^{\perp}$ and $E(K^{\perp}) = \{\xi \in \widehat{R} \mid \widehat{K}\xi = 0\}$.

4. ESSENTIALLY DENSE SUBDIRECT PRODUCTS

At least three concepts have been used to measure how far a subdirect product M of modules $M \subset \Pi\{M_i \mid i \in I\} = T$ is from being their full direct product T. The classical notion that $M \subset T$ be dense in the product (Tychonoff) topology on T is too weak to afford a good description of M. The other two inequivalent and different concepts of an "essential subdirect product $R \subset \prod_I R_i$ of rings" defined in [Go; p. 115] and [Lo; p. 91] fortunately coincide under the additional hypothesis that all the rings $\{R_i \mid i \in I\}$ are right nonsingular. Our subdirect products will be "dense" or "essential" in all of the above senses. A module M is an irredundant subdirect product $M \subset \prod_I M_i$ if the kernel of the natural projection $\prod_I M_i \longrightarrow \Pi\{M_i \mid i \in I \setminus \{k\}\}$ intersects M nontrivially for all k, that is if $M \cap M_k \neq 0$ for all $k \in I$.

4.1. Definition. A subdirect $M \subset \Pi\{M_i \mid i \in I\} \equiv T$ of right R-modules M_i is an *essentially dense* subdirect product if

$$\bigoplus_I (M \cap M_i) << M << T_R.$$

Note that "essentially dense" implies "irredundant".

The next lemma would be false if the nonsingular hypothesis were omitted.

4.2. Lemma. Let $R \subset \Pi\{R_i \mid i \in I\} \equiv T$ be a subdirect product of right nonsingular rings R_i with identity, and let "$\leqq_e$" denote essential right ideals in the rings R_i and T; (R is not assumed to have an identity).

(1) Then the following four conditions are equivalent.

    (a) $R << T_R$.

    (b) $\forall\ i,\ R_i \cap R \leqq_e R_i$.

    (c) $\displaystyle\bigoplus_{i \in I} (R_i \cap R) << T_R$.

    (d) $\exists\ J \leqq_e T,\ J \subseteq R$.

If (1) (a) - (d) hold, then

    (2) $ZR = 0$; and

    (3) R is an irredundant subdirect product of the R_i.

Condition 4.3 (ii) below is one of the hypotheses used in [FL; p.251, Theorem 2] to guarantee irredundancy. We state it for the sake of completeness without any nonsingularity restrictions on the R_i.

4.3. Corollary 1. Suppose that $K \subset \Pi\{R_i \mid i \in I\} = T$ is an essentially dense subdirect product of semiprime rings R_i. Then

    (i) R is semiprime; and for all i.

    (ii) $\{x \mid x \in R_i,\ x(R_i \cap R) = 0\} = \{y \mid y \in R_i,\ (R_i \cap R)y = 0\}$.

4.4. Corollary 2. Suppose that $\{M_i \mid i \in I\}$ are torsion free right R-modules, and that $M \subset \Pi\{M_i \mid i \in I\} = T$ is an essentially dense subdirect product. If each M_i is molecular, discrete, continuous, bottomless, or semiprime, then M_R and T_R are likewise.

The next corollary is not an immediate special case of the previous one.

4.5. Corollary 3. Suppose that $R \subset \Pi\{R_i \mid i \in I\} = T$ is a subdirect product of nonsingular rings R_i so that 4.2 (1) (a) - (d) hold. If each R_i is molecular, discrete, continuous, or bottomless over the ring R_i, then R and T are likewise as right R-modules.

4.6. Proposition. Suppose that $\{A_i \mid i \in I\}$ is a family of nonzero torsion free modules with $\operatorname{Hom}_R(\widehat{A}_i, \widehat{A}_j) = 0$ for all $i \neq j$. Then

    (i) $E(\bigoplus_{i \in I} A_i) = \prod_{i \in I} \widehat{A}_i$.

Take any $0 \neq x_i \in A_i$ and any $C_i < R$ with $x_i^{\perp} \oplus C_i << R,\ i \in I$. Then

    (ii) $Z_2 R + \sum_{i \in I} C_i = Z_2 R \oplus (\bigoplus_{i \in I} C_i) \leq R$;

    (iii) $\forall\ i \neq j \in I,\ C_i C_j = 0$.

Proof. (i) Regard $E(\bigoplus_I A_i) \leq \prod_I \widehat{A}_i$. It suffices to show that for any $0 \neq \xi = (\xi_i) \in \prod_I \widehat{A}_i$, and any $\xi_i \neq 0$, there exists an $r \in R$ with $0 \neq \xi r = \xi_i r \in \widehat{A}_i < \prod_I \widehat{A}_i$. Since A_i

is torsion free there exists $0 \neq C < R$ with $\xi_i^\perp \oplus C \leq R$. Since $\xi_i^\perp <\!\!\not< R$, $C \neq 0$. Suppose that $\xi_j C \neq 0$ for some $j \neq i$. Since $\xi_i^\perp \cap C = 0$, $C \cong \xi_i C$ and the map $\xi_i C \longrightarrow \xi_j C$, $\xi_i r \longrightarrow \xi_j r$, $r \in C$ extends to a nonzero element of $\operatorname{Hom}_R(\widehat{A}_i, \widehat{A}_j)$, a contradiction.

(ii) For each i, since A_i is t.f., firstly, $ZR \subset x_i^\perp$, and secondly, $x_i^\perp < R$ is a right complement by 2.1. Then $Z_2 R \subset x_i^\perp$ by 2.3 (3). Hence $Z_2 R \subseteq \bigcap_I x_i^\perp$.

First suppose that $\sum_I C_i$ is not a direct sum. Then let $2 \leq n$ be minimal such that $C_1 + \cdots + C_n$ is not a direct sum, where $1, \ldots, n \in I$ are distinct. Then there are $0 \neq a_i \in C_i$ such that $a_n = a_1 + \cdots + a_{n-1} \in C_1 \oplus \cdots \oplus C_{n-1}$. Since $x_i^\perp < R$ is a right complement, $(C_i \oplus x_i^\perp)/x_i^\perp <\!\!< R/x_i^\perp$. Thus $\widehat{C}_i \cong E(x_i R)$. Hence $a_n R \hookrightarrow E(x_n R)$ and $a_n R \hookrightarrow E(x_1 R) \oplus \cdots \oplus E(x_{n-1} R)$. If the latter inclusion is followed by the projection maps into the $E(x_i R)$, it then follows that $\operatorname{Hom}_R[E(x_n R), E(x_i R)] \neq 0$ for some $i \leq n-1$, a contradiction. Thus $\sum_I C_i = \bigoplus_I C_i$. Since each $ZC_i = 0$, also $Z(\bigoplus_I C_i) = 0$. Then $Z_2 R \cap (\bigoplus_I C_i) = 0$ because $ZR <\!\!< Z_2 R$.

(iii) Suppose that $zC_i \neq 0$ for some $z \in C_j$. Define an R-map $f : C_j \longrightarrow C_i$ by $fr = zr$ for $r \in C_j$. Since C_i is t.f.; the kernel $\ker f = z^\perp \cap C_j < C_j$ is a right complement (in C_j only, not R). Hence $(z^\perp \cap C_j) \oplus V <\!\!< C_j$ for some $0 \neq V < C_j$. Consequently

$$C_j \supset V \cong zV \subset C_i.$$

Thus V is isomorphic to submodules of $E(x_j R)$ and $E(x_i R)$, thereby contradicting that $\operatorname{Hom}_R[E(x_i R), E(x_j R)] = 0$.

The number of elements $|I|$ of such a rigid system $\{A_i\}_{i \in I}$ of modules as above implies that at least $|R| \geq |I|$. The next corollary might possibly be useful in computing Goldie dimensions of R. It is proof is immediate from [DF; p.299, Theorem 3].

4.7. Corollary 1. Under the assumptions of the previous proposition

$$|I| \leq \sum_{i \in I} \operatorname{Gd} C_i \leq \operatorname{Gd}(R/Z_2 R).$$

If in the previous proposition R is a torsion free semiprime ring in which $\bigoplus C_i <\!\!< R$, then we have a special instance of the next proposition. Note that the right ideals $A_i < R$ below need not be complements.

4.8. Proposition. Suppose that R is a t.f. ring and that $A_i < R$, $i \in I$, are right ideals satisfying the following for all $k \in I$:

(a) $\bigoplus_{i \in I} A_i <\!\!< R$;

(b) $\forall\ i \neq j \in I$, $A_i A_j = 0$; and

(c) $A_k \cap A_k^\perp = 0$.

Define $\widetilde{A}_i = (A_i \oplus A_i^\perp)/A_i^\perp$. Then

(i) $\forall\ J \subseteq I$, $[\bigoplus_{j \in J} A_j]^\perp = [\bigoplus_{i \in I \setminus J} A_j]^- \lhd R$; in particular if $J = \{k\}$, then

(ii) $A_k^\perp = [\bigoplus_{i \in I, i \neq k} A_i]^- \lhd R$; $A_k \oplus A_k^\perp <\!\!< R$; and $\bigcap_{i \in I} A_i^\perp = 0$.

(iii) R is isomorphic to an essentially dense subdirect product of

$$R \rightarrowtail\!\!\!\!\!\twoheadrightarrow \tilde{R} \subset \prod_{i \in I} R/A_i^\perp = T$$

where $A_k \cong \tilde{A}_k \leq_e R/A_k^\perp$, and $\bigoplus_{i \in I} \tilde{A}_i << \tilde{R} << T_R$.

(iv) $\hat{A}_k$ are rings and as rings and as a right R-module

$$\hat{R} \cong \prod_{i \in I} E(R/A_i^\perp) \cong \prod_{i \in I} \hat{A}_i$$

Proof. (i) By (b), $\bigoplus_{I \backslash J} A_i \subseteq [\bigoplus_J A_j]^\perp \leq R$, where the latter is a right complement by 2.1. Thus if the conclusion $[\bigoplus_{I \backslash J} A_i]^- \leq [\bigoplus_J A_j]^\perp$ is proper, then $[\bigoplus_{I \backslash J} A_i]^- <\!\!\not< [\bigoplus_J A_j]^\perp$, and hence $[\bigoplus_{I \backslash J} A_i]^- \oplus D \leq [\bigoplus_J A_j]^\perp$ for some $0 \neq D < R$. From (a) it follows that there exists a $0 \neq d \in D \cap (\bigoplus_I A_i)$ of the form $d = a + b$, $b \in \bigoplus_{I \backslash J} A_i$ and $0 \neq a = a_1 + \cdots + a_n \in A_{j(1)} \oplus \cdots \oplus A_{j(n)} = \bigoplus_J A_j$, $a_k \in A_{j(k)}$, $j(k) \in J$. Thus we get from (b) that $\bigoplus_J A_j b = 0$ and $0 = \bigoplus_J A_j d = \bigoplus_J A_j a = A_{j(1)} a_1 \oplus A_{j(2)} a_2 \oplus A_{j(n)} a_n$, or $a_k \in A_{j(k)} \cap A_{j(k)}^\perp = 0$ for all k, which is a contradiction. Hence $[\bigoplus_{I \backslash J} A_i]^- = [\bigoplus_J A_j]^\perp$. (ii) Note that $(\bigoplus_J A_j) \oplus [\bigoplus_J A_j]^\perp << R$ and $[\bigoplus_J A_j]^\perp = \bigcap_J A_j^\perp$. Hence $(\bigoplus_I A_i)^\perp = \bigcap_I A_i^\perp = 0$ because R is t.f., and for $J = \{k\}$, $A_k \oplus A_k^\perp << R$.

(iii) Since $A_k^\perp < R$ is a right complement and $A_k \oplus A_k^\perp << R$, it follows that also $\tilde{A}_k = (A_k \oplus A_k^\perp)/A_k^\perp << R/A_k^\perp$. Since $\bigcap_{i \in I} A_i^\perp = 0$, the map $R \longrightarrow \tilde{R}$, $r \longrightarrow (r + A_i^\perp)_{i \in I}$, $r \in R$, is an isomorphisms of rings and right R-modules which maps the essential submodule $\bigoplus_I A_i << R$ exactly onto its essential isomorphic image $\bigoplus_I \tilde{A}_i << \tilde{R}$. Now by 4.2 (1)(c), $\tilde{R} <<T_R$.

(iv) Again from $\tilde{A}_k << R/A_k^\perp$ we get that $\hat{A}_k \cong E(R/A_k^\perp)$ as right R-modules. The proof that $\hat{A}_k$ is a ring and that this is a ring isomorphism is omitted. The rest follows from 4.6 (i).

5. FULLY INVARIANT INTRINSIC SUBMODULES

We now combine the classification of modules into types (1.2) and apply the facts about complement closures (section 2) to obtain an internal description of torsion free modules (Theorem I) and rings (Corollary 2 to Theorem I). This section isolates that what can be deduced without semiprimeness hypotheses. Later we apply this as well as the previous sections to obtain a structure theorem for certain semiprime torsion free rings.

The following argument is used repeatedly later on.

5.1. Observation. If $\{A_\gamma \mid \gamma \in \Gamma\}$ is any family of modules, then every nonzero submodule V of $\leq E(\bigoplus\{A_\gamma \mid \gamma \in \Gamma\})$ contains a nonzero submodule which is isomorphic to a cyclic submodule of some single A_γ for some $\gamma \in \Gamma$.

Proof. For $0 \neq \xi \in E(\bigoplus A_\gamma)$ chose $r_0 \in R$ such that $0 \neq \xi r_0 = a_1 + \cdots + a_n \in A_{\gamma(1)} \oplus \cdots \oplus A_{\gamma(n)}$ where all $0 \neq a_i \in A_{\gamma(i)}$, and with the length n minimal. If $z \in a_i^\perp \setminus a_j^\perp$, then $\xi r_0 z \neq 0$ has shorter length. Consequently

$$(\xi r_0)^\perp = a_1^\perp = a_2^\perp = \cdots = a_n^\perp, \text{ and}$$
$$V \supset \xi r_0 R \cong a_1 R \subset A_{\gamma(1)}.$$

Now a Zorn's lemma argument could be used to show that V actually contains an essential direct sum of such nonzero cyclic submodules each of which is isomorphic to a cyclic submodule of some A_γ for various $\gamma \in \Gamma$.

5.2. Consider any one of the following possible six classes of modules: (a) discrete, (b) continuous, (c) continuous molecular, (d) bottomless, (e) $\tau \in \Omega$, and (f) for any fixed arbitrary type $0 \neq [Q] \in \Xi(R)$, the class $\bigcup\{[P] \mid 0 \neq [P] \in \Xi(R), [P] \leq [Q]\} = \{P \mid \exists\, 0 \neq P \hookrightarrow E(\bigoplus\{Q \mid \Gamma\})$ for some set $\Gamma\}$. Then this class is closed under the following operations: (i) nonzero submodules, (ii) injective envelopes of arbitrary direct sums, (iii) isomorphic copies, and (iv) nonzero torsion free homomorphic images (i.e. quotient modules modulo proper complement submodules). It has been proven ([D9; p.62, 3.15 (iii), (iv)]) that (a) - (e) are merely special cases of (f).

5.3. Let M be a t.f. module and $\alpha \in \Xi(R)$. Then
   (1) $E(M_\alpha) = (EM)_\alpha$.

Proof. Since $M_\alpha < EM$, and α is closed under injective hulls, $E(M_\alpha) \leq (EM)_\alpha$. Since $E(M_\alpha)$ is a direct summand of $(EM)_\alpha$ it follows that $E(M_\alpha) = (EM)_\alpha$. Write $\widehat{M}_\alpha = E(M_\alpha) = (EM)_\alpha$.

   (2) $M_\alpha < M$ and $\widehat{M}_\alpha < EM$ are fully invariant right complement submodules.

   (3) In particular, if $ZR = 0$, $M = R_R$, and $\tau \in \Omega$, then $R_\tau \lhd R$ and $R_\tau < R$ is a right complement.

5.4. Theorem I. Suppose that M is a unital torsion free right R-module over any ring R. Let $\rho \neq \tau \in \Omega = \Omega^C \cup \Omega^D$ as in 1.2 and $M_\tau < M$ as in 1.3. There exist $A, B, C, D \leq M$ submodules such that

   (a) A is continuous molecular, B bottomless, C continuous, and D discrete.
   (b) $A = \overline{A}$, $B = \overline{B}$, $C = \overline{C}$, and $D = \overline{D}$ are right complement submodules of M.
   (c) $A \oplus B \oplus D << M$, $C \oplus D << M$, $A \oplus B << C$.
   Then the following hold.
   (i) A, B, C, and D are fully invariant in M; $\mathrm{Hom}_R(\widehat{A}, \widehat{B}) = 0$, and $\mathrm{Hom}_R(\widehat{C}, \widehat{D}) = 0$.
   (ii) UNIQUE: If $A_1 \oplus B_1 << C_1$ and $C_1 \oplus D_1 << M$ satisfy (a), (b), and (c), then $A = A_1$, $B = B_1$, $C = C_1$, and $D = D_1$.
   (iii) $M_\tau < M$ is a fully invariant complement for any $\tau \in \Omega$. (See 1.3.)
   (iv) $\sum_{\tau \in \Omega} M_\tau = \bigoplus_{\tau \in \Omega} M_\tau << A \oplus D;\ \bigoplus_{\tau \in \Omega^C} M_\tau << A;\ \bigoplus_{\tau \in \Omega^D} M_\tau << D$.
   (v) $\mathrm{Hom}_R(\widehat{M}_\tau, \widehat{M}_\rho) = 0$ for any $\tau \neq \rho \in \Omega$.

Proof. (i) and (ii). By Zorn's lemma, let C_0 (D_0) be any maximal direct sum of continuous (discrete) submodules of M. Set $C = \overline{C}_0$ and $D = \overline{D}_0$. Then $C \oplus D << M$. Now select any maximal direct sums A_0 and B_0 of molecular and bottomless submodules of C. Set $A = \overline{A}_0 \leq M$ and $B = \overline{B}_0 \leq M$. Again, $A \oplus B << C$ and $A \oplus B \oplus D << M$. By 5.2 (ii), $\widehat{A}$ is molecular, $\widehat{B}$ bottomless, $\widehat{C}$ continuous, and $\widehat{D}$ discrete. Also, $\widehat{M} = \widehat{A} \oplus \widehat{B} \oplus \widehat{D}$ and $\widehat{C} = \widehat{A} \oplus \widehat{B}$. By 2.6 (iv), $\mathrm{Hom}_R(\widehat{C}, \widehat{D}) = 0$ and $\mathrm{Hom}_R(\widehat{A}, \widehat{B}) = 0$, thus showing that all these submodules of $\widehat{M}$ are fully invariant. Finally, their uniqueness and full invariance in $\widehat{M}$ automatically implies the uniqueness and full invariance of $M \cap \widehat{A} = A$, $M \cap \widehat{B} = B$, $M \cap \widehat{C} = C$, and $M \cap \widehat{D} = D$ in M.

(iii) Every nonzero homomorphism of a uniform module into a torsion free module is a monomorphism. Every nonzero torsion free homomorphic image of an atomic module W is likewise atomic by 2.6 (iv). Hence $M_\tau < M$ is fully invariant and a complement submodule by 5.3 (2).

(iv) If $W_1 \in \tau$ and $W_2 \in \rho$ for $\tau \neq \rho \in \Omega$, then by $[W_1] \neq [W_2]$ are atoms in $\Xi(R)$. It follows from this that W_1 does not contain a nonzero submodule isomorphic to a submodule of W_2. An argument similar to the one used in the proof of 4.6 (ii) shows that the sum in (iv) is direct.

(v) Since M_τ contains an essential direct sum of a subset of τ, it follows by 5.2 (ii) that $[\widehat{M_\tau}] = \tau$. The fact that $\rho \neq \tau \in \Xi(R)$ are atoms implies that no nonzero submodule of $\widehat{M_\tau}$ is isomorphic to a submodule of $\widehat{M_\rho}$. Thus $\mathrm{Hom}_R(\widehat{M_\tau}, \widehat{M_\rho}) = 0$.

5.5. Corollary 1 to Theorem I. For an arbitrary torsion free injective module M, suppose that $M = A \oplus B \oplus D$ and $M_\tau = \widehat{M_\tau}$ are the direct summands of M given by the last theorem. Then

(i) $A \oplus D = E(\bigoplus_{\tau \in \Omega} M_\tau) = \prod_{\tau \in \Omega} M_\tau$;

(ii) $A = \prod_{\tau \in \Omega^C} A_\tau$ and $D = \prod_{\tau \in \Omega^D} D_\tau$.

Proof. (i) This follows from 5.4 (iv) and 4.6 (i).

(ii) For $\tau \in \Omega^C$, $M_\tau = A_\tau$ and similary for $\tau \in \Omega^D$, $M_\tau = D_\tau$.

By use of [D8; p.17, 4.4] it can be shown that in the above corollary D is isomorphic to a dense subdirect product of pairwise isomorphic indecomposable injective modules. It would be interesting to obtain a similar description of A.

5.6. Corollary 2 to Theorem I. For any ring R $(1 \in R)$ with $ZR = 0$, let $A \oplus B \oplus D << C \oplus D << R$ be the right ideals given by the last theorem, and let $\rho \neq \tau \in \Omega$ be arbitrary. Then the following hold:

(i) $A, B, C, D \triangleleft R$; $CD = DC = 0$, $AB = BA = 0$, i.e. the product of any two distinct ideals in any order is zero.

(ii) $A, B, C, D \leq R$ are complement right ideals.

(iii) UNIQUE. A is the unique smallest complement right ideal of R which contains every continuous atomic (or molecular) right ideal. Moreover, A is the sum of all molecular right ideals of R. Parallel statements hold for B, C, and D.

(iv) $R_\tau \triangleleft R$.

(v) $\sum_{\tau \in \Omega} = \bigoplus_{\tau \in \Omega} R_\tau << A \oplus D$; in particular $R_\tau R_\rho = 0$.

(vi) $R_\tau^\perp = [\bigoplus\{R_\rho \mid \tau \neq \rho \in \Omega\} \oplus B]^-$; $R_\tau \oplus R_\tau^\perp << R$.

(v) (a) Conclusions (i) - (v) hold for the ring $\widehat{R} = \widehat{C} \oplus \widehat{D} = \widehat{A} \oplus \widehat{B} \oplus \widehat{D}$ (if every occurence of R, A, B, D, and D is replaced by $\widehat{R}, \widehat{A}, \widehat{B}, \widehat{C}$, and $\widehat{D}$).

(b) $\widehat{A} \oplus \widehat{D} = E(\bigoplus_{\tau \in \Omega} R_\tau) \cong \prod_{\tau \in \Omega} \widehat{R}_\tau$ as a right R-modules and as rings. In particular, $\widehat{R}_\tau$, $\widehat{R}_\rho \lhd \widehat{R}$ with $\widehat{R}_\tau \widehat{R}_\rho = 0$ and

(c) $\widehat{A} = E(\bigoplus_{\tau \in \Omega^C} A_\tau) \cong \prod_{\tau \in \Omega^C} \widehat{R}_\tau$ and $\widehat{D} = E(\bigoplus_{\tau \in \Omega^D} D_\tau) \cong \prod_{\tau \in \Omega^D} \widehat{R}_\tau$ as right R-modules and as rings.

By using Corollary 5.6, it is shown below that every t.f. ring R is a subdirect product $R \subset (R_1 \times R_3) \times R_2$ of a t.f. molecular ring $(R_1 \times R_3)$ and a bottomless ring R_2. More important than the statement of the next corollary is its proof which actually constructs the R_i explicitly from certain unique ideals inside R. Under the additional assumptions that R is semiprime molecular, $R_2 = 0$, and later we will obtain a detailed internal description of such rings as well as a necessary and sufficient characterization.

5.7. Corollary 3 to Theorem I. Any right nonsingular ring R, i.e. with $ZR = 0$ is (isomorphic to) a subdirect product $R \subseteq R_1 \times R_2 \times R_3 \equiv T$ of right nonsingular rings R_i where R_1 is a continuous molecular ring, R_2 is a bottomless ring, and R_3 is a discrete ring with $(R \cap R_1) \oplus (R \cap R_2) \oplus (R \cap R_3) << R << T_R$. Moreover, R is semiprime $\iff$ all R_i are semiprime rings.

Proof. Let $A \oplus B \oplus D << C \oplus D << R$ be as in 5.6. Define $R_1 = R/A^\perp$, $R_2 = R/B^\perp$, $R_3 = R/D^\perp$, $\widetilde{A} = [A \oplus A^\perp]/A^\perp \lhd R_1$, and similarly $\widetilde{B} \lhd R_2$, $\widetilde{D} \lhd R_3$, and $\widetilde{C} \lhd R/C^\perp$. Then $A^\perp \cap B^\perp \cap D^\perp = 0$; $R \cong \widetilde{R}$ under the canonical map $R \rightarrowtail\!\!\!\rightarrow$ $\widetilde{R} \subseteq R_1 \times R_2 \times R_3$; $\widetilde{R}$ is an irredundant subdirect product of the R_i. Since $\widetilde{A} << R/A^\perp$, the latter is first, molecular as a right R-module. From this it can be argued that as a ring in its own right, R_1 is continuous molecular. Similarly, R_2 is bottomless and R_3 discrete. Furthermore $\widetilde{R}$ is a large R-submodule of T as follows

$$A \oplus B \oplus D \cong \widetilde{A} \oplus \widetilde{B} \oplus \widetilde{D} << \widetilde{R} << T_R.$$

By 2.1 and 3.7, the rings R_i are semiprime if R is.

We next show how R can also be represented as a subdirect product of a continuous ring and a discrete, and furthermore that this simpler subdirect product representation fits inside the previous bigger one.

5.8. Construction. Regard $R/C^\perp = R/[A^\perp \cap B^\perp] \subset (R/A^\perp) \times (R/B^\perp)$ as a subring and a right R-submodule via $r + C^\perp \longrightarrow (r + A^\perp, r + B^\perp)$ for $r \in R$. Then our previous ring $\widetilde{R}$ in 5.7 is a subdirect product of the continuous t.f. ring $R/C^\perp$ and the t.f. discrete ring $R/D^\perp$ with

$$C \oplus D \cong \widetilde{C} \oplus \widetilde{D} << \widetilde{R} << \frac{R}{C^\perp} \times \frac{R}{D^\perp} << T_R = R_1 \times R_2 \times R_3.$$

Again, if R is semiprime, so are $R/C^\perp$ and $R/D^\perp$.

The next corollary follows from Proposition 4.8.

5.9. Corollary 4 to Theorem I. Suppose that R is a t.f. molecular ring and $\bigoplus\{R_\tau \mid \tau \in \Omega\} << A \oplus D << R$ are as in 5.6. Define $\widetilde{R}_\tau = (R_\tau \oplus R_\tau^\perp)/R_\tau^\perp \cong R_\tau$. Then R is isomorphic as a ring and right R-module to an essentially dense subdirect product of atomic rings

$$R \rightarrowtail\!\!\!\twoheadrightarrow \tilde{R} \subset \prod_{\tau \in \Omega} R/R_\tau^\perp = T$$

with $\displaystyle\bigoplus_{\tau \in \Omega} \tilde{R}_\tau << \tilde{R} << T_R.$

6. APPLICATIONS TO RINGS

The semiprime and prime module theoretic developments of section 3 are now combined with some of the results of section 4 but mainly with those of section 5 and applied to rings.

6.1. Lemma. For a t.f. semiprime ring R, suppose that $0 \neq P$, $0 \neq Q \leq R$ are right ideals such that their types are comparable, i.e. either $0 \neq [P] \leq [Q] \in \Xi(R)$, or $[Q] \leq [P]$. Then $PQ \neq 0$.

In particular, for any $0 \neq P$, $0 \neq Q \leq R$ such that $P, Q \subseteq R_\tau$ for some $\tau \in \Omega$, $PQ \neq 0$.

Proof. Assume $[Q] \leq [P]$. Then $Q \hookrightarrow E(\bigoplus\{P \mid \Gamma\})$ for some index set Γ. Now 4.1 shows that for any $0 \neq \xi \in Q$, there are $r_0 \in R$ and $a \in P$ such that $0 \neq \xi r_0 R \cong aR$. So far semiprimeness has not been used, but now it is needed to apply 3.5 (with $B = aR$ and $A = \xi r_0 R$) to conclude that $0 \neq aR\xi r_0 R \subset PQ$.

If $P, Q \subseteq R_\tau$, then $0 \neq [P] \leq [R_\tau] = \tau \in \Xi(R)$. But since τ is an atom, necessarily $[P] = \tau = [Q]$ in this case.

In the next theorem, if the atomic right ideal $W < R$ is discrete, then there is a family $\{U_\alpha\}$ of uniform right ideals all of the same type $(\hat{U}_\alpha \cong \hat{U}_\beta$ all $\alpha, \beta)$ such that $\displaystyle\bigoplus_\alpha U_\alpha << W << E(\bigoplus_\alpha U_\alpha)$. As a special case, W itself could be uniform.

6.2. Theorem II. For a semiprime ring R with $ZR = 0$, let $A \oplus B \oplus D << C \oplus D << R$, $\tau \in \Omega$ and $0 \neq R_\tau \lhd R$ be as in 4.8, and let $W < R$ be any atomic right ideal (1.2). Then the following hold.

    (1) (a) W is a prime right R-module.
        (b) $W^\perp \lhd R$ is a prime ideal.
        (c) For any two atomic right ideals $V, W < R$, if $V \cap W \neq 0$, then $\Longrightarrow W^\perp = V^\perp$.
        (d) For any $W \in \tau$, $R_\tau^\perp = W^\perp$. Hence $R_\tau^\perp \lhd R$ is a prime ideal.
    (2) (a) $R_\tau \oplus R_\tau^\perp << R$.
        (b) $R_\tau^\perp$ is a maximal annihilator ideal of R.
    (3) (a) R_τ is a prime right R-module.
        (b) R_τ is prime ring.
        (c) $A, B, C, D \lhd R$ and $R_\tau \lhd R$ are semiprime ideals.
        (d) $A^\perp = (B \oplus D)^-$, $B^\perp = (A \oplus D)^-$, $D^\perp = (A \oplus B)^-$, $C^\perp = D$, $D^\perp = C \lhd R$
all are semiprime ideals.
        (e) $C = (A \oplus B)^-$; $C^\perp = A^\perp \cap B^\perp = D$, $B^\perp \cap D^\perp = A$, $D^\perp \cap A^\perp = B$.

Proof. (1) (a) If W is not a prime R-module, then $w_0 Rt = 0$ for some $0 \neq w_0 \in W$ and some $t \in R \setminus W^\perp$. Therefore $w_1 t \neq 0$ for some $0 \neq w_1 \in W$. Since $0 \neq [w_0 R]$, $0 \neq [w_1 tR] \leq [W] \in \Xi(R)$ and $[W]$ is an atom, $[w_0 R] = [w_1 tR]$. The later implies that there is an embedding $w_0 R \hookrightarrow E(\bigoplus\{w_1 t R \mid \Gamma\})$ for some index set Γ. Now 5.1 shows that $(w_0 r_0)^\perp = (w_1 tr_1)^\perp$ for some $r_0, r_1 \in R$ with $w_1 tr_1 \neq 0$. Then $w_0 r_0 R w_1 tr_1 \subseteq w_0 R t r_1 = 0$. Consequently $R w_1 tr_1 \subseteq (w_0 r_0)^\perp = (w_1 tr_1)^\perp$. But then $(w_1 tr_1) R w_1 tr_1 = 0$ contradicts the semiprimeness of R.

(1) (b) The annihilator ideal of any prime R-module is always a prime ideal of R ([D4; p.160, 1.7(2)])

(1) (c) A module W is prime if and only if for any nonzero submodule, such as $0 \neq V \cap W < W$, $(V \cap W)^\perp = W^\perp$ ([D4; p.159, 1.31 (ii)]). By symmetry, $(V \cap W)^\perp = V^\perp$. Thus $V^\perp = W^\perp$.

(1) (d) Let $V, W < R$ with $V, W \in \tau$. There exist $0 \neq P < V$, $0 \neq Q < W$ with $P \cong Q$ by 5.1. The later together with the fact that V and W are prime modules by (a) guarantees that $V^\perp = P^\perp = Q^\perp = W^\perp$. From $R_\tau = \sum\{W \mid W \leq R, W \in \tau\}$ it follows that for any $0 \neq W \in \tau$, $R_\tau^\perp = \bigcap\{W^\perp \mid W \leq R, W \in \tau\} = W^\perp \lhd R$ is a prime ideal of R by 1(b).

(2) (a) By 5.6 (vi), $R_\tau \oplus R_\tau^\perp \ll R$. (2) (b) If not, then $R_\tau^\perp \underset{\neq}{\subset} I \lhd R$ with $\ell(I)^\perp \neq R = I$ as in 1.1. A technical argument using 3.8 now shows that $\ell(I) \subseteq R_\tau$ and $I \cap R_\tau \neq 0$. But then 6.1 applied to the nonzero atomic right ideals $0 \neq I \cap R_\tau$, $0 \neq \ell(I) \subset R_\tau$ of R yields the contradiction that $\ell(I)(I \cap R_\tau) \neq 0$. Therefore $R_\tau^\perp$ is a maximal annihilator ideal of R.

(3) (a) If not, then $\alpha R t = 0$ for some $0 \neq \alpha \in R_\tau$ and $t \in R \setminus R_\tau^\perp$. Thus $\beta t \neq 0$ for some $\beta \in R_\tau$. Since $0 \neq \alpha R$, $0 \neq \beta t R \subseteq R_\tau$, now 6.1 yields the contradiction that $0 \neq \alpha R \beta t R \subseteq \alpha R t R = 0$.

(3) (b) The proof of 3 (b) is omitted.

(3) (c) In 3.6, take $K = \overline{K} = A, B, C, D$, or R_τ. 3 (d) and (e) follow from 4.8.

The next corollary is the first step towards the solution of the still open problem of finding the real algebraic differences between discrete and continuous rings of the above type R_τ. The conjecture is that the next corollary fails for some continuous rings R_τ with $\tau \in \Omega^C$.

6.3. Corollary to Theorem II. Let R be a torsion free semiprime ring, $\tau \in \Omega^D$, and let $0 \neq I \lhd R_\tau$ be any ideal. Then there exists a $0 \neq J \lhd R$ with $J \subseteq I$ such that $J \subset R_\tau$ is large as a right R_τ-module (and hence also as a right R-module).

Proof. Use of 6.1 shows that for any $V < R_\tau$, $V \ll R_\tau$ if and only if $V \subset R_\tau$ is large as a right R_τ-module. Let $0 \neq \xi \in I \lhd R_\tau$. Since R_τ is a prime ring $0 \neq a\xi b \in (\xi R_\tau)^3$ for some $a, b \in R_\tau$. There is an indexed set of uniform right R-ideals $\{U_\alpha\}$ such that $\bigoplus_\alpha U_\alpha \ll R_\tau$. By 6.1, $U_\alpha a \xi b R \neq 0$ for all α. Consequently $\bigoplus_\alpha U_\alpha a \xi b R \ll R_\tau$. Define $J = R a \xi b R$. Then $J \subseteq I$, $0 \neq J \lhd R$, and $\bigoplus_\alpha U_\alpha a \xi b R \ll J$. Thus $J \ll R_\tau$.

The necessary and sufficient conditions in the next theorem are best possible. That is, counterexamples show that the next theorem simply becomes false if the molecular-atomic hypothesis is omitted (see 7.1). The next theorem does a lot more than merely prove that two sets of conditions are equivalent. It gives a unique representation of certain rings as subdirect products. And it describes exactly how to build this unique representation of R entirely within R from certain unique ideals of R. In the subdirect product representation $R \subset \prod_I R_i$ below, ideally one would like to have $\bigoplus_I R_i \subset R$. This is too much to ask, but we have the next best thing to it – that $\bigoplus_I (R \cap R_i) << R$. In fact, the sum $\bigoplus_I (R \cap R_i)$ turns out to be our previous sum $\bigoplus_\Omega R_\tau$ (see 2 (b)).

6.4. Theorem III. Any ring R with identity satisfies the following.

(1) R is (H1) right nonsingular, i.e. with $ZR = 0$, (H2) semiprime, and (H3) right molecular, $\Longleftrightarrow$ R is a subdirect product $R \subseteq \Pi\{R_i \mid i \in I\} \equiv T$ of rings R_i which are (h1) right nonsinglar, (h2) prime, (h3) right atomic, with (h4) $R << T_R$.

Furthermore if (h1) and (h4) hold then $\bigoplus\{(R \cap R_i) \mid i \in I\} << R$.

(2) Now suppose that R satisfies (H1), (H2), and (H3) above, and let $R_\tau^\perp \oplus R_\tau <<$ R, $\tau \in \Omega$ be as in 5.3. Then

(a) $\bigcap\{R_\tau^\perp \mid \tau \in \Omega\} = 0$; $R \cong \widetilde{R}$ as a ring and as a right R-module, where $\widetilde{R}$ is the following irredundant subdirect product of prime right nonsingular molecular rings $R/R_\tau^\perp$,

$$R \succ\!\!\xrightarrow{\;\cong\;}\!\!>\!> \widetilde{R} \subset \prod_{\tau \in \Omega} R/R_\tau^\perp \equiv T.$$

Moreover, $\widetilde{R}$ is an essential right R-submodule of T, such that

(b) $\bigoplus_{\tau \in \Omega} R_\tau \cong \oplus[(R_\tau^\perp + R_\tau)/R_\tau^\perp] << \widetilde{R} << T_R$.

(3) UNIQUENESS. Suppose that $R \subset \Pi\{R_i \mid i \in I\}$ is any subdirect product representation of R satisfying (1) (h1)-(h4). Then there exists a bijection $f : I \longrightarrow \Omega$ and a ring isomorphism $g : \prod_I R_i \longrightarrow \prod_\Omega R/R_\tau^\perp$ such that $gR_i = R/R_{f(i)}^\perp$ for all $i \in I$.

Proof. The proof of (1) and (2) can be based on 4.8, 4.6, and 4.5 and is omitted. By [L; p.66, Theorem 3.2] and 6.2 (2) (b), it suffices to show that every proper annihilator ideal I is contained in some $R_\tau^\perp$. Assume not. Then $b_\tau I \neq 0$ for some $0 \neq b_\tau \in R_\tau$ for every $\tau \in \Omega$. As in 1.1, $I = \ell(I)^\perp = I^{\perp\perp}$. From $\bigoplus_\Omega R_\tau << R$ we get that there exists $0 \neq \eta = a_1 + \cdots + a_n \in \ell(I) \cap [R_{\tau(1)} \oplus \cdots \oplus R_{\tau(n)}]$, $0 \neq a_i \in R_{\tau(i)}$ such that $\eta^\perp = a_1^\perp = \cdots = a_n^\perp$. Then $\eta R \cong a_1 R$ and $(\eta R)^\perp = (a_1 R)^\perp$. From $\eta R \subset \ell(I)$ we get that $0 \neq b_{\tau(1)} I \subset I = \ell(I)^\perp \subseteq (\eta R)^\perp = (a_1 R)^\perp$, and hence $(a_1 R)b_{\tau(1)} I = 0$. But $0 \neq a_1 R \subset R_{\tau(1)}$ and $0 \neq b_{\tau(1)} I \subset R_{\tau(1)}$ contradict 6.1.

7. EXAMPLES

Some examples are given illustrating Theorems and at the same time these examples show what typical bottomless, discrete, and molecular rings look like.

The next examples show that Theorem III becomes false if the molecular hypotheses (i.e. 6.4 (1); (H3) and (h3)) is omitted. The counterexamples below are commutative, t.f., semiprime, bottomless rings R, which have the further property that every nonzero ideal of R contains a proper zero divisor. Suppose that any one of these rings R was a subdirect product of prime rings R_i with $\bigoplus_I (R_i \cap R) << R << \prod_I R_i = T$ (as in 6.4 (2) (b)). Then the R_i are nonzero prime domains. This is a contradiction, because $0 \neq R_i \cap R \lhd R$.

7.1. Counterexamples. (a) Let D be any commutative domain with identity, and $R = \prod_I^\infty D / \bigoplus_I^\infty D$.

(b) For any infinite set X, let $\mathcal{P}(X)$ be the Boolean ring of all subsets of X with $A \cdot B = A \cap B$ and $A + B = (A \cup B) \setminus A \cap B$ for $A, B \subseteq X$. The finite subsets $\mathcal{F}(X)$ form an ideal of $\mathcal{P}(X)$. Set $R = \mathcal{P}(X)/\mathcal{F}(X)$.

In the next example, R is continuous molecular. It also shows that in Theorem III even a finite subdirect product need not be the whole direct product.

7.2. Example. Let $F = K\{y, z\}$ be the free algebra over any field K in two noncommuting variables y and z. First of all, F is a nonsingular prime continuous atomic ring. Let $\langle y \rangle = FyF \lhd F$ denote the ideal generated by any element $y \in F$. Define R to be the subring

$$R = \{(\alpha + \gamma, \beta + \gamma, \gamma) \mid \alpha \in \langle y \rangle, \ \beta \in \langle z \rangle, \ \gamma \in F\} \subset F \times F \times F = T.$$

Then R is a t.f. continuous molecular semiprime ring. Since F is atomic, it follows that $R_{\tau(1)} = \langle y \rangle \times \{0\} \times \{0\}$, $R_{\tau(2)} = \{0\} \times \{z\} \times \{0\}$, $R_{\tau(3)} = \{0\} \times \{0\} \times \{\langle y \rangle \cap \langle z \rangle\} \lhd R$ are atomic ideals. The poset $\Xi(R)$ is the eight element Boolean lattice of all subsets of $\Omega = \Omega^C = \{\tau(1), \tau(2), \tau(3)\}$.

Let $r = (\alpha + \gamma, \beta + \gamma, \gamma) \in R$ be arbitrary. Cosets modulo $R_{\tau(3)}^{\perp} = \langle y \rangle \times \langle z \rangle \times \{0\}$ have unique coset representatives of the form $r + R_{\tau(3)}^{\perp} = (\gamma, \gamma, \gamma) + R_{\tau(3)}^{\perp}$ where $\gamma \in F$. Hence $R/R_{\tau(3)}^{\perp} \cong F$.

Since $R_{\tau(1)}^{\perp} \subseteq \{0\} \times F \times F$, $R_{\tau(1)}^{\perp} = \{(0, a + b, a) \mid a \in \langle y \rangle, b \in \langle z \rangle\}$. Any $r = (\alpha + \gamma, \beta + \gamma, \gamma) \in R$ is of the form $\gamma = k_0 + zp(z) + h(y)$ where $k_0 \in K$, $p(z) \in K[z]$ and $h(y) \in \langle y \rangle$. Then

$$r + R_{\tau(1)}^{\perp} = r + (0, -\beta - zp(z) - h(y), -h(y)) + R_{\tau(1)}^{\perp}$$
$$= (\alpha + h(y) + k_0 + zp(z), k_0, k_0 + zp(z)) + R_{\tau(1)}^{\perp}.$$

So far, $R/R_{\tau(1)}^{\perp} \hookrightarrow F \times K \times K[z]$, and more importantly, each coset $r + R_{\tau(1)}$ determines a unique element $k_0 + zp(z) \in K[z]$. Thus $r \in R_{\tau(1)}^{\perp}$ iff $\alpha + h(y) = -k_0 - zp(z) \in \langle y \rangle \cap K[z] = 0$, iff $\alpha = -h(y)$, $\gamma = h(y)$, and $r = (0, \beta + h(y), h(y))$. Consequently, the projection $R/R_{\tau(1)}^{\perp} \longrightarrow F$, $r + R_{\tau(1)}^{\perp} \longrightarrow (\alpha + \gamma, 0, 0)$ onto the first component is an isomorphism $R/R_{\tau(1)}^{\perp} \cong F$. Similarly, $R/R_{\tau(2)}^{\perp} \cong F$.

In this example, R is not a direct product of the F's, and in the essential dense subdirect product

$$R_{\tau(1)} \oplus R_{\tau(2)} \oplus R_{\tau(3)} = \langle y \rangle \times \langle z \rangle \times \langle y \rangle \cap \langle z \rangle \ll R \ll F \times F \times F = T,$$

$F \oplus F \oplus F$ is not contained in R.

It is easy to construct examples illustrating the relation $\bigoplus_I (R_i \cap R) \ll R \ll \prod_I R_i = T$ in Theorem III where all $R_i \cap R = R_i$. In the next example, I is infinite, and all $R_i \cap R \leq_e R_i$ are proper inclusions. Moreover, the R_i are not strongly prime.

7.3. A strongly prime ring R is t.f., prime, and atomic.

Proof. Clearly it is prime and torsion free. (See [HL; p.213, II.2].) For any $0 \neq y \in R$, there exist $s_1, \ldots, s_n \in R$ such that $\bigcap_I^n (ys_i)^\perp = 0$. Therefore there is a monomorphism $R \longrightarrow \bigoplus_{i=1}^n ys_iR$, $r \longrightarrow (ys_1r, \ldots, ys_nr)$. Hence $[yR] = [R] \in \Xi(R)$ and R is atomic.

Many examples of strongly prime rings are available, including ones with zero divisors ([HL; p.212, Ex.1, 2, 3] and [GHL]). Next we show that one of the strongly prime rings in [HL; p.212, Ex.1] is a continuous atomic prime t.f. ring with proper zero divisors.

7.4. Example. Let R be the noncommutative polynomial ring $R = Z_2\{x_1, x_2, \ldots, x_i, \ldots\}$ in a countable number of indeterminates over the field $Z_2 = \{0, 1\}$ subject to the relations $x_i x_j x_k = 0$ when $i > j > k$. An arbitrary element $0 \neq c \in R$ is uniquely of the form $c = m_1 + m_2 + \ldots + m_e$ where the $0 \neq m_i$ are monomials. Let $m = x_{i(1)} x_{i(2)} \cdots x_{i(k)}$ be any one of the monomials m_i of highest degree k. Without loss of generality we may assume that $k \geq 1$. Then $(c x_{i(k)} x_1)^\perp \subseteq (m x_{i(k)} x_1)^\perp = 0$. Thus R is a strictly prime ring. Set $0 \neq a = c x_{i(k)} x_1$ and $0 \neq b = c x_{i(k)}^2 x_1$. Then $aR \oplus bR \subseteq cR$ proves that R is continuous.

7.5. Example. For a right vector space V over a division ring F, let $\operatorname{End} V_F$ be the full ring of F-linear operators on V written on the left of V. Set $R_i = \operatorname{End} V_F$ for $i = 1, 2, \ldots$. Let $D \subset \prod_I^\infty R_i = T$ be the diagonal $D = \{(t_i) \mid \exists\, t \in \operatorname{End} V_F, \forall_i\, t_i = t\}$. Let $R_i^0 \lhd R_i$ be the operators with finite dimensional ranges. Define $R = \bigoplus_I^\infty R_i^0 + D$. Then R is an essentially dense subdirect product $\bigoplus(R_i \cap R) << R << \prod_I R_i$ where $R_i \cap R = R_i^0 \neq R_i$. Here $\Omega = \Omega^D = \{\tau(1), \tau(2), \ldots, \tau(i), \ldots\}$ where $R_{\tau(i)} = R_i^0$, and $\Xi(R) = \mathcal{P}(\Omega)$.

For any $y, s_1, \ldots, s_n \in \operatorname{End} V_F$ with $K = \text{kernel } (y)$ of finite codimension, $\bigcap_I^n s_i^{-1} K \neq 0$. Hence there exists a $z \in \operatorname{End}_F V$ with $0 \neq zV \subseteq \bigcap_I^n s_i^{-1} K$ and $ys_i z = 0$ for all $i = 1, \ldots, n$. Thus the three rings $R_i^0 \subset R_i^0 + F \hookrightarrow \operatorname{End} V_F$ are not strongly prime; they are t.f. discrete atomic prime.

At this point we have seen that the class of t.f. discrete atomic prime rings is properly bigger than the discrete strongly prime ones.

REFERENCES

[A1] A. V. Andrunakievic, Prime modules, prime one-sided ideals, and Baer radical, Akademia Nauk, Moldavskoi S.S.R. Izvestiia Seriia Fiziko-Tekh. i Matem. Nauk 2 (1976), 12-17 (In Russian).

[A2] A. V. Andrunakievic, On one-sided completely prime ideals, Izvestija Akad. Nauk Moldavskaja S.S.R., Ser. fiz. - tehn. mat. Nr. 1 (1977), 10-15 (Russian).

[BB] J. Beachy and W. Blair, Rings whose faithful left ideals are cofaithful, Pac. J. Math. 58 (1975), 1-13.

[D1] J. Dauns, One-sided prime ideals, Pac. J. Math. 47 (1973), 401-412.

[D2] J. Dauns, Quotient rings and one sided primes, J. Reine angew. Math. 278/279 (1975), 206-224.

[D3] J. Dauns, Generalized monoform and quasi-injective modules, Pac. J. Math. 66 (1976), 49-65.

[D4] J. Dauns, Prime modules, J. Reine Angew. Math. 298 (1978), 156-181.

[D5] J. Dauns, Prime modules and one-sided ideals, in: Ring Theory and Algebra III, Proc. of the Third Oklahoma Conference (Marcel Dekker, New York, 1979), 41-83.

[D6] J. Dauns, Uniform modules and complements, Houston J. Math. 6 (1980), 31-40.

[D7] J. Dauns, Sums of uniform modules, in: Advances in Noncommutative Ring Theory, Proceedings, Plattsburgh, 1981, Lecture Notes in Math. 951 (Berlin-New York 1982), 68-87.

[D8] J. Dauns, Uniform dimensions and subdirect products, Pac. J. Math. 136 (1987), 1-19.

[D9] J. Dauns, Torsion free modules, Annali di Mat. Pura ed Appl., CLIV (1989), 49-81.

[DF] J. Dauns and L. Fuchs, Infinite Goldie dimensions, J. Alg. 115 (1988), 297-302.

[FL] L. Fuchs and F. Loonstra, Note on irredundant subdirect products, Studia Scientiarum Math. Hungarica 16 (1981), 249-254.

[G] A. W. Goldie, Semiprime rings with maximum condition, Proc. London Math. Soc. 10 (1960), 201-220.

[Go] K. R. Goodearl, Ring Theory (Marcel Dekker, New York, 1976).

[HL] K. R. Goodearl, D. Handelman, and J. Lawrence, Strongly Prime and Completely Torsion-Free Rings, Carleton University Math. Lecture Notes 109, Carleton, Ottawa, 1974.

[H] F. Hansen, On one-sided prime ideals, Pac. J. Math. 58 (1975), 79-85.

[HL] D. Handelman and J. Lawrence, Strongly prime rings, Trans. Amer. Math. Soc. 211 (1975), 209-223.

[Ho] M. Hongan, On strongly prime modules, Math. J. Okayama Univ., 24 (1982), 117-132.

[K1] K. Koh, A note on a certain class of prime rings, Amer. Math. Monthly 72 (1965), 875-877.

[K2] K. Koh, On some characteristic properties of self injective rings, P.A.M.S. 19 (1968), 209-213.

[K3] K. Koh, On one-sided ideals of a prime type, P.A.M.S. 28 (1971), 321-329.

[K4] K. Koh, On one-sided ideals, Canad. Math. Bull. 14 (1971), 244-259.

[K5] K. Koh, Quasi-Simple Modules, in Lectures on Rings and Modules, Lecture Notes in Math. 246 (Springer, Berlin - New York 1972), 349-355.

[L] L. Levy, Unique subdirect sums of prime rings, Trans. Amer. Math. Soc. 106, (1963), 64-76.

[Lo] F. Loonstra, Essential submodules and essential subdirect products, Symposia Math. XXIII (1970), 85-105.

Department of Mathematics
Tulane University
New Orleans, Louisiana 70118

PRIMITIVE IDEALS OF NICE ORE LOCALIZATIONS

Mark L. Teply and Blas Torrecillas*

Abstract. Let R be a ring and let τ be an hereditary torsion theory such that R has the descending chain condition on τ-closed left ideals. Let $\{P_1, P_2, \cdots, P_n\}$ be a link closed set of τ-closed prime ideals and C be a left reversible left Ore set in $C\left(\bigcap_{i=1}^{n} P_i\right)$. If $C^{-1}R$ is left artinian, then τ can be modified to another torsion theory σ such that R has DCC on σ-closed left ideals and the primitive ideals of $C^{-1}R$ are precisely localizations of the σ-closed prime ideals of R. As an application, new information about localization at sets of minimal primes in rings with left Krull dimension and in Noetherian rings is obtained. The condition that the primitive ideals of $C^{-1}R$ are precisely $\{C^{-1}P_1, C^{-1}P_2, \cdots, C^{-1}P_n\}$ is also studied.

1. INTRODUCTION

Considerable work has been done on Ore localizations at sets of prime ideals in Noetherian rings; for example, see [3], [6] and their references. More recently, ideas on linkage (as used in [3]) have been extended to study injective modules and Ore localizations in non-Noetherian rings; for example, see [1], [10], [11], [12] and [13]. These studies build on other work on Ore localizations related to a torsion theory; see [2] for details and references. In this paper we continue the investigation of localizations in a general setting by studying the primitive ideals of Ore localizations arising from Ore sets that lie in $C\left(\bigcap_{i=1}^{n} P_i\right)$, where $\{P_1, P_2, \cdots, P_n\}$ is a collection of primes associated to a torsion theory in a special way. However, one of our results specializes to a generalization of a result of Boyle and Kosler [3, Theorem 3.7] and gives new information about localizations at sets of minimal primes in rings with left Krull dimension.

Let R be a ring with identity element, let τ be an hereditary torsion theory of left R-modules, let $\mathcal{T}$ denote the torsion class, and let $\mathcal{F}$ denote the torsionfree class. (See [2], [4] or [9] for basic results on torsion theory.) For a module M, let $\tau(M)$ denote the largest submodule of M that is in $\mathcal{T}$. A submodule N of M is τ-closed (τ-dense) if $M/N \in \mathcal{F}(M/N \in \mathcal{T})$. If R has the descending chain condition (DCC) on τ-closed left ideals, then R also has the ascending chain condition (ACC) on τ-closed left ideals, and every indecomposable injective module in $\mathcal{F}$ is a direct sum of indecomposable injective modules; moreover, there are only finitely many isomorphism classes of indecomposable injective modules in $\mathcal{F}$ in this case. A nonzero module M is called τ-cocritical if $M \in \mathcal{F}$ but $M/N \in \mathcal{T}$ for every nonzero submodule N of M. An ideal D of R is called τ-primitive if $D = ann_R C$ for some τ-cocritical module C. As proved in [1] and [11], when R has DCC on τ-closed left ideals, there is a one-to-one correspondence between the sets of (1) τ-closed prime ideals of R, (2) isomorphism classes of nonzero indecomposable injective modules in $\mathcal{F}$, and (3) minimal τ-primitive ideals of R. For an indecomposable injective module in $\mathcal{F}$, this correspondence is given by

$$P = ass_R E \leftrightarrow E \leftrightarrow ann_R S_\tau(E) = D \ ,$$

where $ass_R E$ denotes the (necessarily prime) assassinator of E and $S_\tau(E)$ denotes the sum of all τ-cocritical submodules of E. Suppose that under the correspondence, we have

$$P_1 \leftrightarrow E_1 \leftrightarrow D_1 \text{ and } P_2 \leftrightarrow E_2 \leftrightarrow D_2 \ .$$

Then we say $P_1(D_1)$ is linked to $P_2(D_2)$ and we write $P_1 \sim> P_2 (D_1 \sim> D_2)$ if and only if D_2 annihilates a nonzero submodule of $E_1/S_\tau(E_1)$. Other equivalent conditions for D_1 to be linked to D_2 are given in [12]. A set X of τ-closed prime (minimal τ-primitive) ideals is link closed if $A \in X$ and $A \sim> B$ imply $B \in X$. Thus the smallest link closed set containing A consists of A and all A_k such that we can find a "chain" of links:

$$A \sim> A_1 \sim> A_2 \sim> \cdots \sim> A_{k-1} \sim> A_k \ .$$

The concept of link closed plays an important role in the study of left Ore sets. (All left Ore sets in this paper are assumed to be multiplicatively closed and contain 1.) In particular, if C is a left Ore set contained in $C(P_1) = \{x \in R | x + P_1$ is regular in $R/P_1\}$ for some τ-closed prime ideal P_1, then $C \subseteq C(P_k)$ and $C \subseteq C(D_k)$ for every P_k in the smallest link closed set containing P_1 and every minimal τ-primitive ideal D_k corresponding to P_k. Such left Ore sets have been studied in [12].

In view of the previous terminology, we introduce *standard notation that will be used for the remainder of the paper*: R is a ring with 1, τ is a torsion theory of R-modules with DCC on τ-closed left ideals, $\{P_1, P_2, \cdots, P_n\}$ is a link closed set of τ-closed prime ideals, $\{D_1, D_2, \cdots, D_n\}$ is the set of minimal τ-primitive ideals corresponding to $\{P_1, P_2, \cdots, P_n\}$, $D = \bigcap_{i=1}^{n} D_i$, $E = \prod_{i=1}^{n} E(R/P_i)$, $P = \bigcap_{i=1}^{n} P_i$ and C is a left reversible, left Ore set in $C(P)$. We let $\phi : R \to C^{-1}R$ be the canonical map for the localization.

Since the τ-closed prime ideals are playing the same role that the minimal prime ideals play in Noetherian localization, a left Ore localization $C^{-1}R$ is called classical with respect to τ if $C^{-1}R$ is a left artinian ring whose primitive ideals are $\{C^{-1}P_i | i \leq n\}$.

In [12] the left classical localizations (with respect to τ) are characterized. Boyle and Kosler [3] have shown that, for a Noetherian ring, a collection of minimal primes is localizable if and only if it is classically localizable in the sense of Jategaonkar [6]. In Section 2 we obtain a generalization of this result in our setting. In particular, we show in Theorem 2.7 that if $C^{-1}R$ is artinian, then we can slightly modify our torsion theory τ to a new torsion theory σ such that (1) R has DCC on σ-closed left ideals and (2) $C^{-1}R$ is classical with respect to σ. Thus we show that artinian Ore localizations are classical in our setting.

In Section 3 we concentrate on the condition that the link closed set $\{P_1, P_2, \cdots, P_n\}$ gives rise to all the primitive ideals of $C^{-1}R$. The left Ore set C gives rise to a natural torsion theory μ_c such that, for a module M, $\mu_c(M) = \{m \in M | cm = 0$ for some $c \in C\}$. Following the standard notation of [4], we let χ_E denote the largest torsion theory for which E is torsionfree. As observed in [12], $\mu_c \leq \chi_E$ (i.e., $\mu_c(M) \subseteq \chi_E(M)$ for any module M), the χ_E-torsionfree prime ideals are $P_1, P_2, \cdots, P_n$, and the minimal χ_E-coprimitive ideals are $D_1, D_2, \cdots, D_n$. Whenever $(\chi_E(R))^m \subseteq \mu_c(R)$ for some m, we show in Proposition 3.3 that the localization $C^{-1}R$ for which $\{C^{-1}P_1, C^{-1}P_2, \cdots, C^{-1}P_n\}$ is the set of all the primitive ideals of $C^{-1}R$ must have nilpotent Jacobson radical $J(C^{-1}R)$. This allows us to state (again under the same technical hypothesis) two characterizations of the localizations that have $\{C^{-1}P_1, C^{-1}P_2, \cdots, C^{-1}P_n\}$ as their set of primitive ideals.

2. ARTINIAN LOCALIZATIONS

In this section we show in Theorem 2.7 that if $C^{-1}R$ is left artinian then $C^{-1}R$ is essentially classical; that is, we can modify the torsion theory τ to a new torsion theory σ so that $C^{-1}R$ is classical with respect to σ. This can be viewed as a generalization of Boyle and Kosler [3, Theorem 3.7] and gives (in Coro. 2.8) new results for rings with left Krull dimension and for Noetherian rings. (Recall that we continue to use the standard terminology adopted in Section 1.)

To establish our result, we need to prove several lemmas about $\{P_1, P_2, \cdots, P_n\}$ and the prime ideals of $C^{-1}R$.

Lemma 2.1. $P_1, P_2, \cdots, P_n$ are minimal prime ideals of D.

Proof. Assume that Q is a prime ideal and that $D \subseteq Q \subseteq P_i$ for some i. Since Q is prime, $D_j \subseteq Q$ for some j. Since P_i contains only one minimal τ-primitive ideal (see [11, Proposition 4.6]), then $i = j$ and $D_i \subseteq Q \subseteq P_i$. Since P_i is τ-closed and since R/D_i is τ-semicocritical, then P_i/D_i is not essential in R/D_i (see [10] or [11]). Let $(A/D_i) \cap (P_i/D_i) = 0$ with $A/D_i \neq 0$. Then $A \not\subseteq Q$. Since $P_i A \subseteq D_i Q$ and Q is prime, then $P_i \subseteq Q$; so $P_i = Q$.

Lemma 2.2. If Q is a prime ideal that contains $\chi_E(R)$, then Q is either χ_E-dense or χ_E-closed.

Proof. Since $D^m \subseteq \chi_E(R)$ for some m, we must have $D \subseteq Q$. Suppose that Q is not χ_E-closed; then $Q \neq P_i$ for any i. Since Q is prime and $Q \supseteq D$, then $Q \supseteq D_j$ for some j.

If Q/D_j is not essential in R/D_j, then $(Q/D_j) \cap (A/D_j) = 0$ for some χ_E-cocritical module with $ann_R(A/D_j) = P_j$. Hence $Q \subseteq P_j$. By Lemma 2.1 we have $P_j = Q$, which is a contradiction.

Hence Q/D_j is essential in R/D_j. By [10, Proposition 1.1] Q/D_j must be χ_E-dense in R/D_j, and hence Q is χ_E-dense in R.

Lemma 2.3. If R has ACC on μ_c-closed left ideals, then for any two-sided μ_c-closed ideal B of R, $C^{-1}B$ is a two-sided ideal of $C^{-1}R$.

Proof. Let $c \in C$ and consider the chain

$$C^{-1}B \subseteq C^{-1}Bc^{-1} \subseteq C^{-1}Bc^{-2} \subseteq C^{-1}Bc^{-3} \subseteq \cdots$$

of left ideals of $C^{-1}R$. Since μ_c is perfect, $C^{-1}R$ has ACC on left ideals; thus $C^{-1}Bc^{-m} = C^{-1}Bc^{-m-1}$ for some m. Hence $C^{-1}Bc^{-1} \subseteq C^{-1}B$ for any $c \in C$. Consequently $C^{-1}B$ is a two-sided ideal of $C^{-1}R$ whenever B is a two-sided ideal of R.

COROLLARY 2.4. If $C^{-1}R$ is left artinian, then for any two-sided μ_c-closed ideal B of R, $C^{-1}B$ is a two-sided ideal of $C^{-1}R$.

Proof. Since μ_c is perfect, R has DCC on μ_c-closed left ideals. By [4, Prop. 21.14] R has ACC on μ_c-closed left ideals; so the result follows from Lemma 2.3.

Lemma 2.5. Assume that $C^{-1}R$ is left artinian and Q' is a prime ideal of $C^{-1}R$. Then $Q = \phi^{-1}(Q')$ is a prime ideal of R.

Proof. Suppose that $H \supseteq Q$ and $K \supseteq Q$ are ideals of R such that $HK \subseteq Q$. Since Q is μ_c-closed in R, then $(C\ell_{\mu_c}^R(H))K \subseteq Q$. By Corollary 2.4, $C^{-1}(C\ell_{\mu_c}^R(H))$ is a two-sided ideal of $C^{-1}R$. Hence $C^{-1}(C\ell_{\mu_c}^R(H))C^{-1}K \subseteq Q'$. Since Q' is prime, either $C^{-1}(C\ell_{\mu_c}^R(H)) \subseteq Q'$ or $C^{-1}K \subseteq Q'$. Thus either

$$H \subseteq \phi^{-1}(C^{-1}H) \subseteq \phi^{-1}(C^{-1}(C\ell_{\mu_c}^R(H))) \subseteq \phi^{-1}(Q') = Q$$

or else

$$K \subseteq \phi^{-1}(C^{-1}K) \subseteq \phi^{-1}(Q') = Q.$$

Lemma 2.6. Assume that $(\chi_E(R))^m \subseteq \mu_c(R)$ for some m and that $C^{-1}R$ is left artinian. Then for any prime ideal Q' of $C^{-1}R$, $\phi^{-1}(Q')$ is either χ_E-dense or χ_E-closed.

Proof. By Lemma 2.5, $\phi^{-1}(Q')$ is a prime ideal of R. Since $\phi^{-1}(Q')$ is μ_c-closed, then $(\chi_E(R))^m \subseteq \mu_c(R) \subset \phi^{-1}(Q')$, and hence $\chi_E(R) \subseteq \phi^{-1}(Q')$. So Lemma 2.2 now yields the result.

We can now give the main result of the section, which is a torsion-theoretic generalization of [3, Theorem 3.7]. An analog of [3, Theorem 3.7] has been given in [12].

Theorem 2.7. Let R have DCC on τ-closed left ideals, let $\{P_1, P_2, \cdots, P_n\}$ be a link closed set of τ-closed prime ideals and let $E = \prod_{i=1}^{n} E(R/P_i)$. Let C be a left reversible left Ore set in $\bigcap_{i=1}^{n} C(P_i)$ such that $(\chi_E(R))^m \subseteq \mu_c(R)$ for some m and $C^{-1}R$ is left artinian. Then there exists a torsion theory $\sigma \leq \chi_E$ such that:

(1) R has DCC on σ-closed left ideals and

(2) $C^{-1}R$ is a left classical localization with respect to σ.

Proof. Assume that $C^{-1}R$ is not a classical localization with respect to τ. Then there exist primitive ideals $Q_1, Q_2, \cdots, Q_t$ of $C^{-1}R$ such that $Q_j \neq C^{-1}P_i$ for any $i = 1$, $2, \cdots, n$ and $j = 1, 2, \cdots, t$. By Lemma 2.5 $\phi^{-1}(Q_j)$ is a μ_c-closed prime ideal of R. Also $\phi^{-1}(Q_j)$ is always χ_E-dense by Lemma 2.6.

Consider the torsion theory

$$\pi = \Lambda_{j=1}^t \chi(E(R/\phi^{-1}(Q_j)))$$

(see [4] for this standard notation). Since each $\phi^{-1}(Q_j)$ is μ_c-closed, then $\mu_c \leq \pi$. Let $\sigma = \pi \Lambda \chi_E$. Since $\mu_c \leq \pi \Lambda \chi_E = \sigma$, then R has DCC on σ-closed left ideals.

We show that $C \subseteq C(\phi^{-1}(Q_j))$ for any j. Let $c \in C$ and assume that $rc \in Q$ for some $r \in R$, where $Q = \phi^{-1}(Q_j)$. Since c is a unit in $C^{-1}Q$, then $r \in C^{-1}Q$. Write $r = c_1^{-1}x$ with $c_1 \in C$ and $x \in Q$. Then $c_2(c_1r - x) = 0$ for some $c_2 \in C$. Thus $c_2c_1r = c_2x \in Q$. Since Q is μ_c-closed, we obtain $r \in Q$. Thus $c + Q$ if left regular in R/Q. By [5, Lemma 3.33] c is regular in R/Q; i.e., $c \in C(Q)$.

From the preceding paragraph, we conclude that

$$C \subseteq \left(\bigcap_{i=1}^n C(P_i)\right) \cap \left(\bigcap_{j=1}^t C(\phi^{-1}(Q_j))\right).$$

Moreover, since $\{C^{-1}P_1, \cdots, C^{-1}P_n, Q_1, \cdots, Q_t\}$ are all the maximal ideals of $C^{-1}R$, then $\{P_1, \cdots, P_n, \phi^{-1}(Q_1), \cdots, \phi^{-1}(Q_t)\}$ must be a link closed set of σ-closed prime ideals. Thus $C^{-1}R$ is a classical localization with respect to σ.

Specializing Theorem 2.7 to rings with left Krull dimension, we have the following result.

COROLLARY 2.8. Let R be a ring with left Krull dimension, let X be a set of minimal prime ideals of R, and let $C \subseteq \bigcap_{P \in X} C(P)$. Suppose that $(\chi \prod_{P \in X} E(R/P)(R))^m \subseteq \mu_c(R)$ for some m. If $C^{-1}R$ is left artinian, then there exists a set Y of prime ideals such that $X \subseteq Y$ and $\left(\bigcap_{P \in Y} C(P)\right)^{-1} R$ is a classical left localization.

Proof. Let $\tau = \chi(\bigoplus_{P \in X} E(R/P))$ in the notation of [4]. By Theorem 2.7 there exists $\sigma \leq \chi_E$ such that $C^{-1}R$ is classical with respect to σ. Let Y be the set of σ-closed prime ideals of R that correspond to maximal ideals of $C^{-1}R$. Then by [12, Theorems 4.4 and 4.5], $(\bigcap_{P \in Y} C(P))^{-1} R$ is classical.

Since Noetherian rings have Krull dimension, Corollary 2.8 applies to give a new result for Noetherian rings. When $C = \bigcap_{P \in Y} C(P)$ in Corollary 2.8, then $\chi \prod_{r \in x} (E(R/P)) = \mu_c$ and our construction yields $Y = X$; so we recover [3, Theorem 3.7] (see also [12, Section 4]).

3. WHEN $C^{-1}R$ HAS NICE PRIMITIVE IDEALS

$C^{-1}R$ is a classical localization when $C^{-1}R$ is left artinian and $\{C^{-1}P_1, C^{-1}P_2, \cdots, C^{-1}P_n\}$ is the set of primitive ideals of $C^{-1}R$. In Section 2 we examined the condition $C^{-1}R$ is left artinian. We now turn our attention to the condition: $\{C^{-1}P_1, C^{-1}P_2, \cdots, C^{-1}P_n\}$ is the set of primitive ideals of $C^{-1}R$. When $(\chi_E(R))^m \subseteq \mu_c(R)$ for some m, then $C^{-1}P$ is nilpotent and hence we can give two characterizations of the condition (in Proposition 3.4).

To obtain our results, we need two lemmas about the set $\{C^{-1}P_1, C^{-1}P_2, \cdots, C^{-1}P_n\}$.

Lemma 3.1. $C^{-1}P_1, C^{-1}P_2, \cdots, C^{-1}P_n$ are minimal prime ideals of $C^{-1}D$.

Proof. Let K/D be a χ_E-cocritical submodule of R/D; then K/D has a nonzero submodule B/D such that $P_i(B/D) = 0$ for some P_i. Hence $C^{-1}P_i(C^{-1}B/C^{-1}D) = 0$ by [12, Proposition 3.4]. Let Q be a prime ideal with $C^{-1}D \subseteq Q$. Then either $C^{-1}P_i \subseteq Q$ or else $C^{-1}B \subseteq Q$.

If we have $C^{-1}B \subseteq Q$ for every such B, then $Q/C^{-1}D$ contains an essential submodule of $C^{-1}R/C^{-1}D$. But $C^{-1}R/C^{-1}D$ is χ_E-semicocritical; so $Q/C^{-1}D$ is χ_E-dense in $C^{-1}R/C^{-1}D$ by [10, Proposition 1.1]. Since each $C^{-1}P_j$ is χ_E-closed, we cannot have $Q \subseteq C^{-1}P_j$ for any j.

Since $\{C^{-1}P_1, \cdots, C^{-1}P_n\}$ is a set of incomparable prime ideals, it now follows that this is a set of minimal prime ideals of $C^{-1}D$.

Remark. As a consequence of the proof of Lemma 3.1, we have the following result: if Q is a prime ideal of $C^{-1}R$ such that $Q \supseteq C^{-1}D$ and $Q \not\supseteq C^{-1}P_i$ for any i, then Q is χ_E-dense in $C^{-1}R$.

Lemma 3.2. Assume that $(\chi_E(R))^m \subseteq \mu_c(R)$ for some m and that $\{C^{-1}P_1, C^{-1}P_2, \cdots, C^{-1}P_n\}$ are the only primitive ideals of $C^{-1}R$. Then every prime ideal of $C^{-1}R$ is maximal, and the prime and Jacobson radicals of $C^{-1}R$ coincide.

Proof. Let Q' be a prime ideal of $C^{-1}R$, and let $Q = \phi^{-1}(Q')$. Then $Q' = C^{-1}Q$, Q is a two-sided ideal of R, Q is μ_c-closed, and $Q \supseteq \mu_c(R)$. Let $K = C\ell^R_{\chi_E}(Q)$.

If $K \neq R$, then $C^{-1}K$ is a two-sided χ_E-closed ideal of $C^{-1}R$, which is contained in a maximal ideal M of $C^{-1}R$. Since M is primitive, then $M = C^{-1}P_i$ for some i by hypothesis. Thus $Q' \subseteq C^{-1}K \subseteq C^{-1}P_i$. But by [10, Theorem 3.2], there exists k such that $(C^{-1}D)^{mk} = C^{-1}D^{mk} \subseteq C^{-1}(\chi_E(R))^m \subseteq C^{-1}\mu_c(R) = 0$. Thus $C^{-1}D \subseteq Q'$; so $Q' = C^{-1}P_i$ by Lemma 3.1.

If $K = R$, then Q is χ_E-dense in R and hence $Q' = C^{-1}Q$ is χ_E-dense in $C^{-1}R$. Since Q' is contained in a maximal (and hence primitive) ideal, then $Q' \subseteq C^{-1}P_i$ for some i, which contradicts the fact that $C^{-1}P_i$ is χ_E-closed in $C^{-1}R$.

We are now ready to consider the nilpotence of $J(C^{-1}R)$.

PROPOSITION 3.3.

(1) If $(\chi_E(R))^m \subseteq \mu_c(R)$ for some m, then nil subrings of $C^{-1}R$ are nilpotent.

(2) If $(\chi_E(R))^m \subseteq \mu_c(R)$ for some m and if $\{C^{-1}P_1, C^{-1}P_2, \cdots, C^{-1}P_n\}$ are all the primitive ideals of $C^{-1}R$, then $C^{-1}P = J(C^{-1}R)$, and $C^{-1}P$ is nilpotent.

Proof.

(1) By [12, Proposition 3.4] $C^{-1}\chi_E(R)$ is a χ_E-closed two-sided ideal of $C^{-1}R$, and $C^{-1}R$ has ACC on χ_E-closed submodules. Thus $C^{-1}R/C^{-1}\chi_E(R)$ is a left Goldie ring. So by Lanski's Theorem [7], nil subrings of $C^{-1}R/C^{-1}\chi_E(R)$ are nilpotent. Let N be a nil subring of $C^{-1}R$. Then $(N + C^{-1}\chi_E(R))/C^{-1}\chi_E(R)$ is nilpotent by the previous paragraph; say $N^k \subseteq C^{-1}\chi_E(R)$. Thus $N^{km} \subseteq (C^{-1}\chi_E(R))^m = C^{-1}(\chi_E(R))^m \subseteq C^{-1}\mu_c(R) = 0$.

(2) Combine Lemma 3.2, [8, Theorem 4.21], and part (1) of this proposition.

Using Proposition 3.3 and the hypothesis that $(\chi_E(R))^m \subseteq \mu_c(R)$ for some m, we can now characterize the condition: $\{C^{-1}P_1, C^{-1}P_2, \cdots, C^{-1}P_n\}$ is the set of primitive ideals of $C^{-1}R$.

PROPOSITION 3.4. Let R have DCC on τ-closed left ideals, and let $\{P_1, P_2, \cdots, P_n\}$ be a link closed set of τ-closed prime ideals. Let C be a left reversible left Ore set in $\bigcap_{i=1}^{n} C(P_i)$ such that $(\chi_E(R))^m \subseteq \mu_c(R)$ for some m. Then the following statements are equivalent.

(1) $\{C^{-1}P_1, C^{-1}P_2, \cdots, C^{-1}P_n\}$ is the set of all primitive ideals of $C^{-1}R$.

(2) $C^{-1}R/C^{-1}P_i$ is a simple ring for each $i = 1, 2, \cdots, n$, and $C^{-1}P = \bigcap_{i=1}^{n} C^{-1}P_i$ is nilpotent.

(3) (a) $C^{-1}R/J(C^{-1}R)$ is a direct sum of simple rings;
    (b) $C^{-1}R/J(C^{-1}R)$ is χ_E-torsionfree; and
    (c) the prime and Jacobson radicals of $C^{-1}R$ coincide.

Proof.

(1) $\Rightarrow$ (2). If $C^{-1}R/C^{-1}P_i$ were not simple, then $C^{-1}R$ would have a maximal (and hence primitive) ideal M containing $C^{-1}P_i$. By (1) $M = C^{-1}P_j$ for some j. But $C^{-1}P_j \supsetneq C^{-1}P_i$ implies that $P_j \supsetneq P_i$, which contradicts the incomparability of $\{P_1, P_2, \cdots, P_n\}$[11]. That $C^{-1}P$ is nilpotent is Proposition 3.3 (2).

(2) $\Rightarrow$ (3). From (2) it is clear that $C^{-1}P = \bigcap_{i=1}^{n} C^{-1}P_i$ is both the prime and Jacobson radical of $C^{-1}R$; so (a) and (c) are clear, and (b) follows easily from [12, Proposition 3.4].

(3) $\Rightarrow$ (1). By (a) and (c), every prime ideal of $C^{-1}R$ is maximal. Thus we only need to show that each maximal ideal of $C^{-1}R$ is some $C^{-1}P_i$. Let M be a maximal ideal of $C^{-1}R$. By (a) $C^{-1}R/M$ is isomorphic to a direct summand of $C^{-1}R/J(C^{-1}R)$; so M is χ_E-closed in $C^{-1}R$ by (b). Thus $C^{-1}R/M$ contains a χ_E-cocritical submodule K/M such that $ann_R(K/M) = P_i$ for some i. Thus $C^{-1}P_iK \subseteq M$. Since M is prime, $C^{-1}P_i \subseteq M$. Since the prime ideals of $C^{-1}R$ are maximal, then $C^{-1}P_i = M$ as desired.

Remarks.

(A) In [12, Example 4.6] $C^{-1}R/C^{-1}P_i$ is a field for each P_i, $C^{-1}R/J(C^{-1}R)$ is a direct sum of fields, and $J(C^{-1}R)$ is nilpotent, but $\{C^{-1}P_1, C^{-1}P_2, \cdots, C^{-1}P_n\}$ is a proper subset of the primitive ideals of $C^{-1}R$. This shows that condition 3(b) is necessary in Proposition 3.4.

(B) In the proof of Proposition 3.4, the hypothesis that $(\chi_E(R))^m \subseteq \mu_c(R)$ for some m is not used in the implications (2) $\Rightarrow$ (3) and (3) $\Rightarrow$ (1).

REFERENCES

* Supported by grant PS88-0108 from DGICYT.

1. T. Albu, F-semicocritical modules, F-primitive ideals and prime ideals, Rev. Roumaine Math. Pures Appl. **31** (1986) 449-459.

2. T. Albu and C. Năstăsescu, Relative finiteness in module theory, Texts in Pure and Appl. Math. 84, Marcel Dekker, New York, 1984.

3. A.K. Boyle and K.A. Kosler, Localization at collections of minimal prime, J. Algebra **119** (1988) 147-161.

4. J.S. Golan, Torsion theories, Pitman monographs and surveys in pure and appl. Math., Longman Scientific Publishing, London, 1986.

5. K.R. Goodearl, Ring theory, Texts in Pure and Appl. Math. 33, Marcel Dekker, New York, 1976.

6. A.V. Jategaonkar, Localizations in Noetherian rings, London Math. Soc. Lecture Notes 98, Cambridge University Press, London, 1985.

7. C. Lanski, Nil subrings of Goldie rings are nilpotent, Can. J. Math. **21** (1969) 904-907.

8. N.H. McCoy, The theory of rings, Macmillan, New York, 1964.

9. B. Stenström, Rings and modules of quotients, Lecture Notes in Math. 237, Springer-Verlag, Berlin, 1971.

10. M.L. Teply, Modules semicocritical with respect to a torsion theory and their applications, Israel J. Math. **54** (1986) 181-200.

11. M.L. Teply, Semicocritical modules, University of Murcia Publications, Murcia, Spain, 1987.

12. M.L. Teply, Links, Ore sets and classical localizations, preprint, 1989.

13. B. Torrecillas, Links between closed prime ideals, preprint.

Department of Mathematical Sciences
University of Wisconsin-Milwaukee
Milwaukee, WI 53201
USA

and

Departamento de Algebra
Facultad de Ciencias
Campus de Almeria
Universidad de Granada
04071 Almeria
SPAIN

FILTERED CARTAN MATRICES FOR ARTINIAN RINGS
OF LOW LOEWY LENGTH*

Kent R. Fuller and Birge Zimmermann-Huisgen

1. Introduction and Elementary Observations

Given a (basic) left artinian ring R with Jacobson radical J and primitive idempotents $e_1, \cdots, e_n$, let c_{ij} be the number of simple composition factors isomorphic to $S_i = Re_i/Je_i$ occurring in a composition series of Re_j. The $n \times n$ matrix $C = [c_{ij}] \in M_n(\mathbf{Z})$ is called the *(left) Cartan matrix* of R. In 1954 Eilenberg showed that finiteness of the left global dimension of R entails det $C = \pm 1$[2], and recently his conclusion was sharpened to det $C = 1$ for large classes of rings, *e.g.*, for rings of left global dimension at most two, left serial rings, positively graded rings, and rings with $J^3 = 0$ (see [9], [1], [8], [5]). Moreover, it turned out that for a left serial ring R, the converse holds as well: namely det $C = 1$ implies finiteness of the left global dimension of R (see [1]). However, the serial rings appear to be rather exceptional in permitting such a painless global dimension test: in fact, even for $J^2 = 0$ and $n = 2$, the condition det $C = 1$ need not force the left global dimension to be finite.

In establishing some of the above results, it proved extremely useful to replace the classical Cartan matrix by certain "filtered" matrices which store considerably more information on the ring while requiring approximately the same computational input. Here we mainly consider the *J-filtered Cartan matrix* $\hat{C} = [\hat{c}_{ij}] \in M_n(\mathbf{Z}[T])$ (filtered with respect to the "radical filtration"), which is defined as follows: $\hat{c}_{ij} = \sum c_{ij}^{(k)} T^k$, where $c_{ij}^{(k)}$ is the multiplicity of S_i in $J^k e_j/J^{k+1} e_j$, in other words, in addition to counting simple composition factors S_i in Re_j, the polynomial $\hat{c}_{ij}$ keeps track of the "radical layers" in which they occur.

Thus the following question imposes itself: How much more accurately does the condition det $\hat{C} = 1$ indicate finiteness of the global dimension?

On the positive side we show that in case $J^2 = 0$, this condition is equivalent to finiteness of the global dimension. On the negative side we provide an example of rings R and S, both of Loewy length 3, which have the same filtered Cartan matrix, while l gl dim R $< \infty$ and l gl dim S $= \infty$. Concerning the range between these extreme situations: in a few rather specialized cases we illustrate how the value of det $\hat{C}$, combined with some additional homological information on the ring, can make it a matter of elementary combinatorics to pin down the projective resolutions of the simple modules. So, while the class of rings for which the condition det $\hat{C} = 1$ by itself guarantees finiteness of the global dimension appears to be rather narrow, in many situations $\hat{C}$ can be a very useful supplementary tool in answering homological questions.

Throughout R will denote an associative left artinian ring with identity, and $e_1, \cdots, e_n$ will be a complete set of primitive idempotents in R. Since it will not affect the generality of our results to assume that R is basic (meaning $Re_i \not\cong Re_j$ for $i \neq j$), we will do so.

If $P \in M_n(\mathbf{Z})$ is a permutation matrix, *i.e.*, if P is obtained from the identity matrix by permuting rows and columns, then, given any $A \in M_n(\mathbf{Z}[T])$, we call the matrix $P^{-1}AP$ *permutation equivalent* to A. Thus the matrices permutation equivalent to C, resp. $\hat{C}$, are just the Cartan matrices, resp. filtered Cartan matrices, of R corresponding to the various indexing of our set of primitive idempotents.

In the following easy observations p dim M will stand for the projective dimension of a left R-module M.

Observation 1. If C (or, equivalently, $\hat{C}$) is permutation equivalent to an $n \times n$ upper triangular matrix, all of whose diagonal entries are equal to 1, then l gl dim R $\leq n-1$.

Indeed, this follows from a straightforward induction on the hypothesis "p dim M $\leq k-1$ if all composition factors of M belong to $\{S_1, \cdots, S_k\}$".

Observation 2. Let $\hat{C} = [\hat{c}_{ij}]$ be the filtered Cartan matrix as above. If det $\hat{C} = 1$, then

$$\hat{c}_{ii} = 1 + \sum_{k \geq 2} c_{ii}^{(k)} T^k \text{ for all } i.$$

Indeed, the constant term of the polynomial $\hat{c}_{ij}$ is clearly 1 if $i = j$ and 0 otherwise. Thus our observation is an immediate consequence of the equality

$$\det \hat{C} = \sum_{\pi \in S_n} sgn(\pi) \hat{c}_{1,\pi(1)} \cdot \quad \cdots \quad \cdot \hat{c}_{n,\pi(n)} .$$

Observation 3. Denote the columns of $\hat{C}$ by $\hat{c}_1, \cdots, \hat{c}_n$, and let $1 \leq k \leq n$. Then p dim $S_k \leq 1$ if and only if there are nonnegative integers $t_1, \cdots, t_{k-1}, t_{k+1}, \cdots, t_n$ such that

$$(k \rightarrow) \begin{bmatrix} 0 \\ \vdots \\ 0 \\ 1 \\ 0 \\ \vdots \\ 0 \end{bmatrix} = \hat{c}_k - \sum_{i \neq k} t_i T \hat{c}_i .$$

2. The Case $J^2 = 0$

According to Jans and Nakayama [6], finiteness of ℓ gl dim R implies ℓ gl dim $R \leq n-1$ when $J^2 = 0$. We supplement this result by the following easy test for finiteness of the global dimension.

Theorem 4. If $J^2 = 0$, the following statements are equivalent:

(1) $\det \hat{C} = 1$.

(2) l gl dim R $< \infty$.

The proof uses the following lemma, which is linear algebra of the most elementary ilk. We nonetheless sketch a proof, since no reference seems to be available.

Lemma 5. Suppose that D is a nilpotent integral (or real) $n \times n$ matrix that has only nonnegative entries. Then D is permutation equivalent to a strictly upper triangular matrix.

Proof. Write $D = [d_{ij}]$, and observe that an easy induction will complete the proof once we have checked that D must contain a zero row.

To verify this last claim, assume the existence of a map $\tau : \{1, \cdots, n\} \to \{1, \cdots, n\}$ such that $d_{i,\tau(i)} \neq 0$ for all i. Since the d_{ij} are nonnegative, it follows that for any positive integer m, the entry in position $(i, t^m(i))$ of D^m is larger than or equal to the following (positive) product

$$d_{i,\tau(i)} \cdot d_{\tau(i),\tau^2(i)} \cdot \quad \cdots \quad \cdot d_{\tau^{m-1}(i),\tau^m(i)} \; ;$$

in particular, we obtain $D^m \neq 0$ for all m, which contradicts our hypothesis.

Proof of Theorem 4. (2) => (1) follows from [5, Corollary to Theorem B].

Now suppose that (1) holds. Observation 2 yields a matrix $[a_{ij}]$ of nonnegative integers such that $\hat{c}_{ii} = a_{ii} = 1$ for all i and $\hat{c}_{ij} = a_{ij}T$ for $i \neq j$. If for each permutation $\pi \in S_n$ we set

$$m(\pi) = card\{i | 1 \leq i \leq n \text{ and } \pi(i) \neq i\} ,$$

then clearly $m(\pi) = 0$ or $m(\pi) \geq 2$, whence

$$1 = \det \hat{C} = 1 + \sum_{k=2}^{n} \left(\sum_{m(\pi)=k} sgn(\pi)a_{1,\pi(1)} \cdot \quad \cdots \quad \cdot a_{n,\pi(n)} \right) T^k .$$

Consequently,

$$\sum_{m(\pi)=k} sgn(\pi)a_{1,\pi(1)} \cdot \quad \cdots \quad \cdot a_{n,\pi(n)} = 0$$

for all $k \geq 2$. Computing the characteristic polynomial $\chi(\lambda)$ of $\hat{C}$, we thus obtain

$$\chi(\lambda) = (1-\lambda)^n + \sum_{k=2}^{n} \left(\sum_{m(\pi)=k} sgn(\pi)a_{1,\pi(1)} \cdot \quad \cdots \quad \cdot a_{n,\pi(n)} \right) T^k (1-\lambda)^{n-k} = (1-\lambda)^n$$

This implies $(\hat{C} - I_n)^n = 0$, where I_n is the $n \times n$ identity matrix, and evaluating at $T = 1$, we obtain $(C - I_n)^n = 0$. At this point, we can either appeal to [6, Proposition 10] or else use Lemma 5 and Observation 1 to deduce that l gl dim R $\leq n - 1$.

3. The Case $J^3 = 0$

If $J^2 = 0$, then, while det $C = 1$ need not imply finiteness of the global dimension, the classic Cartan matrix C itself does contain enough information to distinguish between finite and infinite global dimension (see [6] and [1, Remark 10]). Any hope that $\hat{C}$ might so serve for $J^3 = 0$ is dashed by the diagram algebras R and S derived from the diagrams

(see [4, Theorem 3.2]). Both algebras have filtered Cartan matrix

$$\hat{C} = \begin{pmatrix} 1 & T & 0 \\ T & 1 + T^2 & T \\ 0 & T & 1 + T^2 \end{pmatrix}$$

One easily checks, e.g., by applying [4, Theorem 2.5], that l gl dim R $= 4$, while l gl dim S $= \infty$. We remark that one can also realize these examples as monomial algebras as follows: $R = F\Gamma / <\rho>$ and $S = F\Gamma / <\sigma>$, where F is a field, Γ the quiver

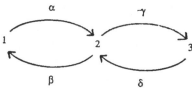

$\rho = \{\beta\alpha, \delta\gamma, \beta\delta, \gamma\alpha\}$, and $\sigma = \{\beta\alpha, \alpha\beta, \beta\delta, \gamma\alpha\}$.

These examples involve a minimal number of primitive idempotents, as the following very specialized result shows. In fact, if $n = 2$, projective resolutions of the two simple modules can simply be read off the filtered Cartan matrix.

Proposition 6. If $J^3 = 0$ and $n = 2$, then the following statements are equivalent:

(1) det $\hat{C} = 1$.
(2) l gl dim R $< \infty$.
(3) l gl dim R ≤ 2.

Proof. For (2) $\Rightarrow$ (1) see [5, Corollary to Theorem B].
(1) $\Rightarrow$ (3). Using Observation 2, we deduce from (1) that

$$\hat{C} = \begin{pmatrix} 1 + aT^2 & bT + cT^2 \\ dT + eT^2 & 1 + fT^2 \end{pmatrix}$$

and

$$(a + f)T^2 + afT^4 = bdT^2 + (be + cd)T^3 + ecT^4 \ ,$$

where $a, \cdots, f$ are nonnegative integers.

In case $a + f = 0$, that is $a = f = 0$, we derive from $bd = be = cd = ec = 0$ that $\hat{C}$ is upper triangular. By Observation 1, R is hereditary in that case.

In case $a + f \neq 0$, say $f \neq 0$, we obtain $b \neq 0$ and $d \neq 0$, whence $e = c = a = 0$ and $f = bd$. This yields $Je_1 \cong S_2^d$ and $Je_2 \cong (Re_1)^b$, whence l gl dim $R = 2$. The case where $a \neq 0$ is symmetric.

By contrast, note that for $J^4 = 0$ the filtered Cartan matrix need not even store the information necessary to deal with the case of two idempotents. In fact, there are diagram algebras R and S, both derived from

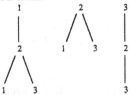

(see [4]), such that gl dim $R = 3$, while gl dim $S = \infty$. Moreover, Kirkman and Kuzmanovich [7] constructed a series of finite dimensional algebras with $J^4 = 0$ and two primitive idempotents whose global dimensions are finite but not bounded above.

Our final proposition shows that, while Proposition 6 cannot be extended to the situation where $J^3 = 0$ and $n = 3$, there is an extension provided that one of the simple modules is known to have projective dimension at most 1. Note that the filtered Cartan matrix $\hat{C}$ by itself reveals whether this additional hypothesis is satisfied (Observation 3).

Proposition 7. Suppose that $J^3 = 0$, $n = 3$, and that R has a simple left module of projective dimension at most 1. Then the following statements are equivalent:

(1) det $\hat{C} = 1$.
(2) l gl dim R $< \infty$.
(3) l gl dim R ≤ 4.

Proof. (1) => (3). Assume that p dim $S_1 \leq 1$.

If $Je_1 = 0$, then

$$R \cong \begin{pmatrix} e_1 Re_1 & e_1 J(e_2 + e_3) \\ 0 & (e_2 + e_3)R(e_2 + e_3) \end{pmatrix} ;$$

so by [3, Corollary 3.6], l gl dim R $\leq 1 + $ l gl dim $(e_2 + e_3)R(e_2 + e_3)$. Since

$$\hat{C} = \begin{pmatrix} 1 & * \\ 0 & \hat{B} \end{pmatrix}$$

our hypothesis implies det $\hat{B} = 1$. Note moreover that $\hat{B}$ is the filtered Cartan matrix of $(e_2 + e_3)R(e_2 + e_3)$. Therefore, Proposition 6 yields l gl dim R ≤ 3.

If p dim $S_1 = 1$, then $Je_1 \cong (Re_2)^{t(2)} \oplus (Re_3)^{t(3)}$ with $t(2) + t(3) > 0$; we may assume that $t(2) > 0$, which implies that Re_2 has Loewy length at most 2. Denoting the columns of $\hat{C}$ by $\hat{c}_1$, $\hat{c}_2$, $\hat{c}_3$, we can restate the above isomorphism as follows:

$$\hat{c}_1 = \begin{bmatrix} 1 \\ 0 \\ 0 \end{bmatrix} + t(2)T\hat{c}_2 + t(3)T\hat{c}_3 \ .$$

Thus $1 = \det \hat{C} = \det D$, where

$$D = \begin{pmatrix} 1 & aT & cT + dT^2 \\ 0 & 1 & eT + fT^2 \\ 0 & Bt & 1 + gT^2 \end{pmatrix}$$

has the same second and third columns as $\hat{C}$. In particular, we have $g = be$ and $bf = 0$.

If $b = 0$, then $Je_2 \cong S_1^a$, and hence p dim $S_2 \leq 2$. Since moreover $g = be = 0$, we see that all simple composition factors of Je_3 are isomorphic to S_1 or S_2, which shows p dim $S_3 \leq 3$.

Now suppose $b \neq 0$. As we saw above, this entails $f = 0$. Moreover, due to $Je_2 \cong S_1^a \oplus S_3^b$, we obtain p dim $S_2 \leq \max(2, \text{p dim } S_3 + 1)$, whence it suffices to show that p dim $S_3 \leq 3$.

If, in addition, we assume $t(3) \neq 0$, then the Loewy length of Re_3 is also bounded above by 2, yielding $d = g = 0$; since $g = be$, we infer $e = 0$, which shows that $Je_3 \cong S_1^c$ and p dim $S_3 \leq 2$.

If, on the other hand, $t(3) = 0$, we obtain $Je_1 / J^2 e_1 \cong S_2^{t(2)}$. Considering a projective cover

$$0 \to K \to (Re_1)^c \oplus (Re_2)^e \overset{\phi}{\to} Je_3 \to 0$$

of Je_3, we infer that

$$K = (Je_1)^c \oplus (K \cap (Je_2)^e) \ ,$$

because $J^2 e_3$ has no summands isomorphic to S_2 (indeed, $f = 0$). In particular, $J^2 e_3$ is an epimorphic image of $(Je_2)^e \cong S_1^{ae} \oplus S_3^{be}$ under ϕ. In view of $g = be$, the S_3-homogeneous component of $J^2 e_3$ is isomorphic to that of $(Je_2)^e$, whence

$$K \cong (Je_1)^c \oplus S_1^{ae-d} \ .$$

But this shows p dim $K \leq 1$, and thus p dim $S_3 \leq 3$ as required.

(3) => (2) is trivial, and (2) => (1) follows from [5, Corollary to Theorem B].

In conclusion, we check that the bound on l gl dim R established in Proposition 7 is sharp. Indeed, if R is any diagram algebra obtained from

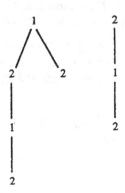

then p dim $S_1 = 1$ and gl dim $R = 4$.

References

* The research of the authors was partially supported by grants from the National Security Agency and the National Science Foundation, respectively.

1. W.D. Burgess, K.R. Fuller, E.R. Voss and B. Zimmermann-Huisgan, The Cartan matrix as an indicator of finite global dimension for artinian rings. Proc. Amer. Math. Soc. **86** (1985) 157-165.

2. S. Eilenberg, Algebras of cohomologically finite dimension. Comment. Math. Helv. **28** (1954) 310-319.

3. R. Fossum, P. Griffith and I. Reiten, Trivial extensions of abelian categories and applications to rings: an expository account, in Ring Theory (R. Gordon, Ed.), Academic Press, New York & London, 1972, pp. 125-151.

4. K.R. Fuller, Algebras for diagrams. J. Pure Appl. Algebra **48** (1987) 23-37.

5. K.R. Fuller and B. Zimmermann-Huisgen, On the generalized Nakayama conjecture and Cartan determinant problem. Trans. Amer. Math. Soc. **294** (1986) 679-691.

6. J.P. Jans and T. Nakayama, On the dimension of modules and algebras. VII. Nagoya Math. J. **11** (1957) 67-76.

7. E. Kirkman and J. Kuzmanovich, Algebras with large homological dimensions. Preprint.

8. G. Wilson, The Cartan map on categories of graded modules. J. Algebra **85** (1983) 390-398.

9. D. Zacharia, On the Cartan matrix of an Artin algebra of global dimension two. J. Algebra **82** (1983) 353-357.

Department of Mathematics
University of Iowa
Iowa City, Iowa 52242

and

Department of Mathematics
University of California
Santa Barbara, California 93106

MORITA CONTEXTS

Philippe Loustaunau and Jay Shapiro

1. DEFINITION AND EXAMPLES

1.1 A *Morita context* is a set (R, V, W, S) and two maps φ and ψ, where R and S are rings, V is a left R-module and a right S-module, W is a left S-module and a right R-module. The maps $\varphi : V \otimes_S W \to R$ and $\psi : W \otimes_R V \to S$ are bilinear. These maps satisfy the associativity conditions that are required to make

$$T = \begin{pmatrix} R & V \\ W & S \end{pmatrix}$$

a ring; namely

$$Id_V \otimes_S \psi = \varphi \otimes_R Id_V \quad \text{and} \quad Id_W \otimes_R \varphi = \psi \otimes_S Id_W \ .$$

T is called the ring of the Morita context.

For $v \in V$ and $w \in W$, we will write vw for $\varphi(v \otimes w)$, and wv for $\psi(w \otimes v)$. Similarly, $Im\varphi$ will be denoted by VW, and $Im\psi$ by WV. These two-sided ideals are called the *trace* ideals of the context. For further reference note that, by the associativity conditions mentioned above, we have $(vw)v' = v(wv')$ and $(wv)w' = w(vw')$, for all $v, v' \in V$ and $w, w' \in W$. Thus, $V(WV) = (VW)V$ and $W(VW) = (WV)W$.

Examples of Morita contexts are abundant and often appear in disguise in the literature. We now list some examples, some of which will be referred to later.

1.2 Let W be a right R-module. Define V to be $Hom_R(W, R)$ and S to be $End_R W$. Then (R, V, W, S) is a Morita context with maps $\varphi : V \otimes_S W \to R$ given by $\varphi(f \otimes w) = f(w)$, and $\psi : W \otimes_R V \to S$ given by $\psi(w \otimes f)(w') = wf(w')$. In particular, if W_R is a progenerator, then R and S are Morita equivalent.

1.3 Let I be a right ideal of R and J be a left ideal of R. Let S be any subring such that $IJ \subseteq S \subseteq I \cap J$. Then (R, J, I, S) is a Morita context, where the maps are the ordinary multiplication in R. As a special case, we have that (R, Re, eR, eRe) is a Morita context, where e is an idempotent.

1.4 Let (R, V, W, S) be a Morita context and T be the ring of the Morita context. Let $e = \begin{pmatrix} 1 & 0 \\ 0 & 0 \end{pmatrix} \in T$. Then clearly eTe is isomorphic to R and $(1-e)T(1-e)$ is isomorphic to S. Conversely, if T is any ring containing an idempotent element e, then we can construct a Morita context using $R = eTe$, $S = (1-e)T(1-e)$, $V = eT(1-e)$ and $W = (1-e)Te$. Moreover, the original ring T is isomorphic to the ring of this Morita context (compare with Example 1.3).

1.5 Let I be a right ideal of a ring R and let S be a subring of R that contains I as a two-sided ideal. Then S is called a *subidealizer* of I in R, and (R, R, I, S) is a Morita context.

1.6 Let G be a finite group of automorphisms acting on R. The *fixed ring* R^G is defined to be the subring $\{x \in R : x^g = x \text{ for all } g \in G\}$. The *skew group ring* $R * G$ is the set of all formal sums $\sum_{g \in G} r_g g$, $r_g \in R$. Addition is componentwise and multiplication is defined distributively by the formula $rg \cdot sh = rs^{g^{-1}} gh$, for $r, s \in R$ and $g, h \in G$. Clearly R is a left and right R^G-module. R can also be viewed as a left or right $R * G$-module as follows: for any $x = \sum_{g \in G} r_g g \in R * G$ and $r \in R$, define $x \cdot r = \sum_{g \in G} r_g r^{g^{-1}}$ and $r \cdot x = \sum_{g \in G} (rr_g)^g$. Then $(R^G, R, R, R * G)$ is a Morita context. The map $\psi : R \otimes_{R^G} R \to R * G$ is defined by $\psi(x \otimes y) = \sum_{g \in G} xy^{g^{-1}} g$, and the map $\varphi : R \otimes_{R * G} R \to R^G$ is defined by $\varphi(x \otimes y) = \sum_{g \in G} (xy)^g$.

2. HISTORICAL SURVEY

2.1 Morita contexts were introduced in 1958 by Morita [Mo]. The definition was suggested by the study of contravariant functors D_1 and D_2 between $Mod - R$ and $Mod - S$ satisfying $D_1 D_2 = Id_{Mod-R}$ and $D_2 D_1 = Id_{Mod-S}$. (He then proved that these functors were Hom-functors.) Ten years later, Bass [B2], in his book K-*Theory*, mentioned Morita contexts under the name of Pre-Equivalence Data. His definition arose naturally from the study of covariant functors D_1' and D_2' between $Mod - R$ and $Mod - S$ that also satisfy $D_1' D_2' = Id_{Mod-R}$ and $D_2' D_1' = Id_{Mod-S}$. (He then proved that these functors were Tensor functors.)

Hom and Tensor functors between $Mod - R$ and $Mod - S$, where R and S are in a Morita context, have been extensively studied. In [Kt2] and [Mb] it was shown that the functors $Hom_S(V, -) : Mod - S \to Mod - R$, and $Hom_R(W, -) : Mod - R \to Mod - S$ induce an equivalence between the quotient categories $\mathcal{U}_R$ of $Mod - R$ and $\mathcal{U}_S$ of $Mod - S$, where $\mathcal{U}_R$ (resp. $\mathcal{U}_S$) is the full subcategory of $Mod - R(Mod - S)$ consisting of all modules L for which the natural map $L \to Hom_R(VW, L)$ (resp. $L \to Hom_S(WV, L)$) is bijective. Later on, in [KO], the above result was dualized using the functors $- \otimes_R V : Mod - R \to Mod - S$, and $- \otimes_S W : Mod - S \to Mod - R$ to obtain an equivalence between the "co-quotient" categories $\mathcal{C}_R$ of $Mod - R$ and $\mathcal{C}_S$ of $Mod - S$, where $\mathcal{C}_R$ (resp. $\mathcal{C}_S$) is the full subcategory of $Mod - R$ (resp. $Mod - S$) consisting of all modules L for which the natural map $L \to L \otimes_R VW (resp. L \to L \oplus_S WV)$ is bijective. Recently, in [NW2], the natural transformation λ from $- \otimes_S W$ to $Hom_S(V, -)$ was shown to have an epi-mono factorization which has an intermediate functor that is an equivalence between the subcategory of trace-accessible, trace-torsion-free submodules. See also [HF], [HL], [Kt1], [Kt3] and [LH] for related results.

2.2 Morita contexts were soon found to be a very useful concept and a unifying tool. For instance, in 1962, Bass used Morita contexts to prove the Wedderburn's Theorems on the structure of simple rings [B1]. In that paper, Example 1.3 (above) plays an important role. It is, in fact, a generalization of an example used by Kaplansky for obtaining the structure theory of primitive rings with minimal ideals (see [Jc], where Morita contexts are used in a disguised fashion). In 1971, Amitsur [Am] used Morita contexts to prove Goldie's Theorem on the ring of quotients of semi-simple rings ([Gd1] and [Gd2]), to prove Wedderburn's structure Theorem for semi-simple artinian rings, and to study the ring of endomorphisms of a module W_R. In particular, he obtained

Zelmanowitz's result [Z] on the structure of $Hom_R(W,W)$ if W_R is torsionless, finitely generated, and R is prime or semi-prime. He also obtained the fact that rings of endomorphisms of primitive rings are primitive (see [Gu] and [Po]). In fact, Morita contexts are used explicitly or implicitly in a large number of papers related to the study of $Hom_R(W,W)$. For example, [An], [Cu], [CRT], [HZ], [Jr1], [Kh1], [Kh2], [Kt1], [Sd], [W] and [Z].

2.3 Morita contexts have been most suitable for the study of transfer of properties from R to S. An example is the study of radicals. In [Am], it was shown that $VN(S)W \subseteq N(R)$, where $N(-)$ is the lower radical, the locally nilpotent radical, the Jacobson radical or the nil radical (if the nil radical of R contains all left (or right) nil ideals). This result was extended in [Sn] to a more general class of radicals. For results related to the radical, see [Jg1], [Jg2] and [NW1].

Another example is the study of the quotient rings of R and S. In [Mb], Müller gives a one-to-one correspondence between those hereditary torsion theories on $Mod - R$ whose filters contain VW, and those torsion theories on $Mod - S$ whose filters contain WV. This correspondence gives rise to an equivalence between the corresponding quotient categories (see 2.2 above). If Q is the quotient functor associated with a torsion theory whose filter contains VW, and if Q' is the corresponding quotient functor in $Mod - S$, then Müller showed that $Q'(S) \widetilde{} End_{Q(R)}(Q(W))$, and $Q(R) \widetilde{} End_{Q(S)}(Q(V))$. Under the hypotheses that $_RV$, V_S, $_SW$, W_R are faithful and that the "products" VW and WV are faithful (*i.e.*, $vW = 0$ implies $v = 0$ and three analogous implications), there is a strong relationship between R and S. For example, under these assumptions, if S has a semi-simple artinian maximal left and right ring of quotients, and if dim $V_S < \infty$, then R has a semi-simple artinian maximal left and right ring of quotients. See also [H], [HL], [HZ], [Ka3] and [LH] for related results.

2.4 The ring $T = \left(\begin{smallmatrix} R & V \\ W & S \end{smallmatrix} \right)$ of the Morita context has also been studied. For example, Sands [Sn] explicitly gives $N(T)$, where N belongs to a general class of radicals (which includes the Jacobson and the nil radicals).

The various rings of quotients of T have also been investigated. Stenström [St] computed the maximal ring of quotients of T when VW and WV are zero. These assumptions were weakened later in [Sk1]: there, the only assumptions are that $_SW_R$ be faithful, and that $vW = 0$ implies $v = 0$, or that VW and WV be nilpotent. Finally, M. Müller [Mm] dropped all restrictions and gave a specific description of the various rings of quotients of T. She did that by studying the injective modules over T and the injective hulls of arbitrary T-modules.

2.5 There is a more general concept than the one of a Morita context. It is the concept of *semi-trivial extensions of rings*. Let A be a ring and M and A-A-bimodule with a bilinear map $\phi : M \otimes_A M \to A$ satisfying $m_1\phi(m_2 \otimes m_3) = \phi(m_1 \otimes m_2)m_3$. The semi-trivial extension $A \times_\phi M$ of A by ϕ is the ring whose underlying set is the cartesian product $A \times M$ with addition componentwise and multiplication given by

$$(a_1, m_1)(a_2, m_2) = (a_1 a_2 + \phi(m_1 \otimes m_2),\ m_1 a_2 + a_1 m_2) \ .$$

f (R, V, W, S) is a Morita context, then by letting

$$A = R \oplus S, \ M = (V \otimes_S W) \oplus (W \otimes_R V) \ \text{and} \ \ \phi = (\varphi, \psi) \,,$$

ve obtain a ring isomorphism between the ring T of the Morita context and $A \times_\phi M$. Note also that $A \times_\phi M$ always embeds in the ring T of the Morita context (R, M, M, R). These extensions have been recently studied in [Ga], [Pa], [Sk2] and [Sk3].

2.6 Finally, for more information of Morita contexts and related studies, see [Am], [Co], [F], [Gl], [Hu], [Ka1], [Ka2], [Kt3], [Mb], [McCR] and [Sw].

3. LOCALIZATION IN A MORITA CONTEXT

As mentioned in Section 2.3, Morita contexts are very suitable to study the transfer of properties from R to S. In the present section, we illustrate this by studying the transfer of the second layer condition and the transfer of right link closed sets from R to S. These properties are necessary to determine classical sets of prime ideals (which allow a well-behaved Ore localization. For example, see [Br] and [Jt3]). The first three results in this section (3.1, 3.2, 3.3) can be found in [LS1].

We now review some definitions of torsion theories as applied to localization at a prime ideal. For further details, see, for example, [Gl].

Let P be a prime ideal of the ring R. An R-module is called *P-Torsion* (resp. *P-Torsion free*) if M is torsion (resp. torsion free) with respect to the torsion theory cogenerated by $E(R/P)$, the injective hull of R/P; *i.e.*, $Hom_R(M, E(R/P)) = 0$ (resp. M embeds in a direct product of copies of $E(R/P)$). A submodule N of M is called *P-dense* in M if M/N is P-torsion.

There are two, possibly different, versions of the second layer condition. Either version can be used to find localizable sets of prime ideals in a right Noetherian ring (see [Jt3]).

Let M be a uniform module over a right Noetherian ring R and let P be the associated prime ideal of M. Then M is called *P-tame* if M is P-torsion free. In general a module is called P-tame if every uniform submodule is P-tame. A prime ideal P of R satisfies the *second layer condition-1* (SLC-1) if there is no finitely generated P-tame uniform module M whose annihilator is equal to a prime ideal strictly smaller that P (this condition was first introduced in [Jt3], see also [Be]). A prime ideal P satisfies the *second layer condition-2* (SLC-2) if, given any short exact sequence of uniform modules,

$$0 \to L \to M \to N \to 0 \,, \quad (*)$$

where M is P-tame and $L = ann_M(P)$ (the set of elements of M annihilated by P), then N must be tame. The prime ideal Q is *right linked* to P if there is a sequence like $(*)$, where all modules are tame and $ass(N) = Q$. This definition of a link is not as originally given in [Jt2] but was proved to be equivalent in [Go].

A set X of prime ideals of R is said to satisfy a second layer condition if each member of X satisfies the condition. Finally, R is said to satisfy a second layer condition if the set of prime ideals of R, denoted $Spec(R)$, satisfies the condition.

A set X of prime ideals is called *right linked closed* if whenever $P \in X$ and Q is right linked to P, then $Q \in X$. If X is a right linked closed set which satisfies SLC-2, then it satisfies SLC-1 [McCR]. While if R is two-sided Noetherian, then it follows from Jategaonkar's Main Lemma [Jt2] that a prime ideal satisfying SLC-1 must satisfy SLC-2.

The way we are going to relate R and S is by "passing" through T; *i.e.*, we are going to show that the properties between prime ideals of R ascend to T and then descend to S. The tools that we will be using are the Hom functors between the different module categories.

In 1.4 above, we have seen that a Morita context (R, V, W, S) can be viewed as the Morita context $(eTe, eT(1-e), (1-e)Te, (1-e)T(1-e))$, where T is the ring $\begin{pmatrix} R & V \\ W & S \end{pmatrix}$ and $e = \begin{pmatrix} 1 & 0 \\ 0 & 0 \end{pmatrix}$. This is the approach we are going to take, since it makes the arguments clearer.

So we assume that T is a right Noetherian ring and that e is an idempotent of T. We let R be eTe and S be $(1-e)T(1-e)$. Then R and S are right Noetherian. Also, there is a lattice bijection between $Spec_e(T) = \{$prime ideals of T which do not contain $e\}$ and $Spec(R)$, given by $P \to ePe$ (with inverse map $P \to \hat{P} = \{t \in T : eTtTe \subseteq P\}$). Finally, the functor $Hom_T(eT, -) : Mod - T \to Mod - R$ is naturally equivalent to multiplication by e.

3.1 Ascent of SLC-1. Let $P \in Spec_e(T)$ such that $ePe = P$ satisfies SLC-1. Then P satisfies SLC-1.

Sketch of the Proof (see [LS1]). We first prove that if M is a P-tame T-module, then Me is a P-tame R-module. For $M = T/P$, it is enough to show that $(1-e)(T/P)e$ is P-tame when $1 - e \notin P$. So we can assume that $P = 0$. If $0 \neq w \in (1-e)Te$, then $(1-e)w = w$ and thus $eT(1-e)w$ is a non-zero subset of R. Now each element of $eT(1-e)$ defines an $R - map$ from $(1-e)Te$ to R; so there is an embedding of $(1-e)Te$ into a direct product of R indexed by $eT(1-e)$. The general case follows from the fact that if M is a uniform P-tame module, then it has a submodule N isomorphic to T/P, and Ne is essential in Me.

Finally, to prove the result, assume to the contrary that there exists a finitely generated P-tame T-module M whose annihilator is equal to P', a prime ideal of T strictly contained in P. Then $P' \in Spec_e(T)$ and we can assume that $P' = 0$. Me is a faithful, finitely generated R-module. But Me is P-tame and $P \neq 0$. This contradicts our assumption on P.

To study the descent of the second layer condition from T to S, we will need the functor $Hom_S(T(1-e), -) : Mod - S \to Mod - T$. This functor is not exact in general (unless $T(1-e)T = T$ in which case T and S are Morita equivalent). To simplify notation, we will denote the functor $Hom_S(T(1-e), -)$ by $(-)^*$.

Note that, if M is an S-module, then $(M*)(1-e) \approx M$.

Recall that the *first layer* of a uniform module M is $ann_M(P)$, where P is the associated prime ideal of M.

3.2 Descent of SLC-2. Let P be a prime ideal of S and let $\mathcal{P}$ be the corresponding prime ideal of T. Suppose that $\mathcal{P}$ belongs to a set Y which has SLC-2 and satisfies the condition that whenever $\mathcal{Q}$ is in Y and $\mathcal{Q}'$ is right linked to $\mathcal{Q}$, then either $\mathcal{Q}'$ is in Y or $e \in \mathcal{Q}'$. Then P satisfies SLC-2.

Sketch of the Proof (see [LS1]). Suppose to the contrary that there is an exact sequence

$$0 \to L \to M \to N \to 0$$

of uniform S-modules, such that M is P-tame, L is the first layer of M, and N is not tame; *i.e.*, N is torsion over its associated prime ideal. We get the following exact sequence

$$0 \to L* \to M* \to N* .$$

By [LS1, Lemma 2.2], $M*$ is a uniform $\mathcal{P}$-tame module. Let A be the first layer of $M*$. Then $A \subseteq L*$ and $A(1-e) = L*(1-e)$. Now, let D be a finitely generated submodule of $M*$ such that the image of D in $M*/L*$ is not killed by $1-e$ (note that $(M*/L*)(1-e) \approx N$). Let $A_0 = A \cap D$ and let $B = D \cap L*$. Then $A_0 \subseteq B \subseteq D$ and $(B/A_0)(1-e) = 0$. Thus

$$0 \neq (D/A_0)(1-e) \approx (D/B)(1-e) \subseteq (M*/L*)(1-e) \approx M/L \approx N .$$

Also D/A_0 is tame.

If one of the associated prime ideals of D/A_0 is in $\mathrm{Spec}_{1-e}(T)$, then N is tame, which is a contradiction. Therefore, we can assume that every associated prime ideal of D/A_0 contains $1-e$. Let U be an essential submodule of D/A_0 which is a direct sum of uniform modules U_i, $i = 1, 2, \ldots, n$. Each prime ideal associated to a U_i is right linked to $\mathcal{P}$ and does not contain e. Therefore, each of these prime ideals is in Y, an so has SLC-2. Let $A_1 \subseteq D$ be the preimage of the sum of the first layers of the $U_i's$. If $\hat{L}_i$ is the first layer of $E(U_i)$, then it follows that D/A_1 embeds in $\oplus \left(E(U_i)/\hat{L}_i \right)$ and hence D/A_1 is tame. Also, it can be shown, as in [LS1, Theorem 2.3], that $A_1 \subseteq B$. We can continue this process, constructing an ascending chain $A_0 \subset A_1 \subset \ldots \subset D$ such that, for each i, all the associated prime ideals of D/A_i contain $1-e$ and so, as before, each of these prime ideals has SLC-2. It can be shown that, for each i, $A_i \subseteq B$.

Since D/A_0 is finitely generated, eventually $A_n = B$ for some n. Thus D/A_n is tame. If all the associated prime ideals of D/A_n contain $1-e$, then we can construct an A_{n+1} strictly containing A_n, which is a contradiction. Thus D/A_n has a tame submodule whose associated prime ideal does not contain $1-e$. Hence N contains a tame submodule. This is a contradiction to our assumption on N.

Note. If P' is right linked to P (in S), then there is a sequence of links from $\mathcal{Q}$ to $\mathcal{P}$ (in T), where $\mathcal{Q}$ corresponds to P', and $\mathcal{P}$ to P. (See also the proof of 3.5 below.)

3.3 Corollary. Let T be two-sided Noetherian. If $T = TeT$ and if R satisfies the second layer condition (the two conditions are equivalent in this case), then S satisfies the second layer condition.

We now turn our attention to the transfer of link closed sets from R to S. In view of the note above, links down in S correspond to a series of links (or long links) up in T. Also one can have links in T between prime ideals of $\text{Spec}_e(T)$ that contain $1 - e$ and prime ideals of $\text{Spec}_{1-e}(T)$ that contain e (see [LS1]). This indicates that in transferring link closed sets from R to S, we may have to "add" prime ideals as we will see in 3.5 below. First, we see that links in T transfer to links in R.

3.4 Transfer of Links from T to R. Let P and Q be in $\text{Spec}_e(T)$ such that Q is right linked to P. Then $eQe = Q$ is right linked to $ePe = P$.

Proof. Let $0 \to L \to M \to N \to 0$ be the exact sequence linking Q to P, where all modules are uniform and tame, L is the first layer of M, and $ass(N) = ann(N) = Q$. We then obtain the following sequence of R-modules
$$0 \to Le \to Me \to Ne \to 0 .$$

All modules are tame and uniform by [LS, Proposition 1.1].

Note that $(Ne)^*$ is isomorphic to $Hom_T(Te \otimes_R eT, N)$, which contains $Hom_T(TeT, N)$. Also from the exact sequence
$$0 \to TeT \to T \to T/TeT \to 0$$

we see that there is a homomorphism from $N \approx Hom_T(T, N)$ to $Hom_T(TeT, N)$. This map is, in fact, one to one, since $Hom_T(T/TeT, N) = 0$ (recall that e is not in $Q = ass(N)$). Thus we can view N as a T-submodule of $(Ne)^*$.

Since $NQ = 0$, it is clear that $ass_R(Ne) \subseteq Q$. In fact, we now show that these are equal. Let X be an R-submodule of Ne. Then $X* \subseteq (Ne)^*$. Now let $Y = X*\cap N$. Then $ann_T(X*) \subseteq ann_T(Y) \subseteq Q$. Furthermore $ann_R(X)$ is contained in $ann_T(X*)$, for if ete is in $ann_R(X)$ and $f : Te \to X$ is in $X* = Hom_R(Te, X)$, then $f \cdot ete(Te) = f(eteTe) = f(e)eteTe = 0$. Thus $ann_R(X)$ is contained in Q, and hence in Q. Therefore, $ass_R(Ne)$ contains Q, which gives equality.

Now let A be the first layer of Me. Then clearly, Le is contained in A. Again we show equality. Let a be in A. Then $aePe = aP = 0$. If $aeP \neq 0$, then aeP is a non-zero T-submodule of M killed by e, which is a contradiction to the fact that M is P-tame and e is not in P. Thus $aeP = 0$, which means that $a = ae$ is in Le. So we have equality.

From all this we can conclude that Q is right linked to P.

Note that if P is a prime ideal of R that contains $eT(1 - e)Te$ (the ideal VW in the Morita context notation), then the prime ideal P of T corresponding to P is such that $(1 - e)P(1 - e) = S$. So, when we talk about the set X' of prime ideals of S corresponding to X, a set of prime ideals of R, we exclude S from X'.

Remark. It follows from 3.1, 3.2 and 3.4 that if T is two-sided Noetherian and if X is a right link closed set in Spec (R) which has the second layer condition, then the corresponding set X' of prime ideals of S satisfies the second layer condition. While X' may not be link closed, our next result gives the best candidate for a link closure of X'.

3.5 Transfer of Link Closed Sets from R to S. Let T be two-sided Noetherian. Let X be a right link closed set of prime ideals of R that satisfies the second layer condition (the two conditions are equivalent in this case). Let X' be the corresponding set of prime ideals of S. If Y is the right link closure of the set of prime ideals of S that contain $(1-e)TeT(1-e)$ (in the notation of a Morita context, this is the ideal WV), then $X' \cup Y$ is a right link closed set in S.

Proof. It is enough to show that if Q' is right linked to P', where P' is in X', then Q' is in $X' \cup Y$. Clearly we can assume that Q' is not in Y. Now let the following exact sequence of S-modules link Q' to P'

$$O \to L \to M \to N \to O \ .$$

In particular, M is P'-tame, L is the first layer of M and N is Q'-tame, and we can assume that N has annihilator Q'. We then obtain the following sequence of T-modules

$$O \to L* \to M* \to N* \ .$$

Let P and Q be the ideals of T corresponding to P' and Q'. Then $M*$ is a uniform P-tame T-module by [LS1, Lemma 2.2]. Since Q' is not in Y, Q does not contain e and hence corresponds to some Q in R. Let A be the first layer of $M*$. As in [LS1, Theorem 2.3], it is not difficult to show that A is contained in $L*$ and that $A(1-e) = L*(1-e)$. In particular $(M*/A)(1-e)$ is equal to $(M*/L*)(1-e)$, which is equal to N. Since P' is in X', it corresponds to a prime ideal P in X, and hence P (in T) satisfies the second layer condition (by 3.1). Therefore, the uniform submodules of $M*/A$ are tame.

Let D be a finitely generated submodule of $M*$ such that the image of D in $M*/L*$ is not killed by $1-e$ (it is possible to find such a D, since $(M*/L*)(1-e) \approx N \neq 0$). Let $A_0 = A \cap D$ and let $B = D \cap L*$. As in [LS1, Theorem 2.3], $A_0 \subseteq B \subseteq D$ and $(B/A_0)(1-e) = 0$. Thus $(D/A_0)(1-e)$ is isomorphic to $(D/B)(1-e)$, which is a non-zero submodule of N by our choice of D. Since $M*/A$ is tame, D/A_0 is also tame. Note that the associated prime ideals of D/A_0 are right linked to P. If one of the associated prime ideals of D/A_0 is in $\mathrm{Spec}_{1-e}(T)$, then it must be Q (by 3.4). Therefore, Q is right linked to P. Since e is not in Q, Q is right linked to P (in R) by 3.4. Therefore, Q is in X, and hence Q' is in X'. Thus we can assume that every associated prime ideal of D/A_0 contains $1-e$, and, therefore, belongs to $\mathrm{Spec}_e(T)$. In particular, every associated prime ideal of D/A_0 corresponds to a prime ideal of R that is right linked to P, by 3.4. Hence every associated prime ideal of D/A_0 satisfies the second layer condition.

Let U be a submodule of D/A_0 that is both essential and a direct sum of uniform submodules U_i, $i = 1, 2, \ldots, n$. Let L_i be the first layer of U_i and let $A_1 \subseteq D$ be the preimage of $\sum L_i$. As noted before, the prime ideal associated to U_i is right linked to P and satisfies the second layer condition, for $i = 1, 2, \ldots, n$. As in 3.2, it can be shown that D/A_1 is tame.

We can continue this process by constructing an ascending sequence

$$A_0 \subset A_1 \subset \ldots \subset D$$

such that, for each i, all the associated prime ideals of D/A_i contain $1 - e$. Each of these prime ideals correspond to a prime ideal in R, and there is a sequence of links between each of these prime ideals (in R) ending with P, and hence they all belong to X. Therefore, the associated prime ideals of D/A_i satisfy the second layer condition (by 3.1). Also, for each i, A_i/A_{i-1} is killed by $1 - e$, and since $1 - e$ is idempotent, A_i/A_0 is killed by $1 - e$. Therefore, $A_i \subseteq B$ as in [LS1, Theorem 2.3].

Since D/A_0 is finitely generated, eventually $A_n = B$ for some n. Thus D/A_n is tame, since its associated prime ideals all satisfy the second layer condition. If all the associated prime ideals of D/A_n contain $1 - e$, then we can construct A_{n+1} strictly containing A_n, which is a contradiction to $A_n = B$. Thus D/A_n has an associated prime ideal which is in $\mathrm{Spec}_{1-e}(T)$. This prime ideal must be $\mathcal{Q}$ (by 3.4). Since $\mathcal{Q}$ corresponds to Q in R, there is a sequence of links from Q to P. Therefore, Q is in X, and Q' is in X'. This completes the proof.

3.6 Corollary. Let T be two-sided Noetherian such that $TeT = T$. Let X be a right link closed set of prime ideals of R satisfying the second layer condition. Let X' be the corresponding set of prime ideals of S. Then X' is a right link closed set in S.

Proof. Since $TeT = T$, the set Y as defined in 3.5 is empty, and the result follows from 3.5.

In the next corollary we use the definition of a *right classical* set of prime ideals as defined in [Jt2]. Furthermore, as proved in [Jt2, Theorem 7.1.5 and Proposition 7.2.4], a finite set X of prime ideals is right classical if and only if X is right link closed, satisfies SLC-2, and is a pairwise incomparable set.

3.7 Transfer of Right Classical Sets. Let T be two-sided Noetherian and suppose that $TeT = T$. If X is a finite classical set of prime ideals of R, then the corresponding set X' of prime ideals of S is also classical.

Proof. By 3.6, X' is a right link closed set. By 3.1 and 3.2, X' satisfies the second layer condition. Since the prime ideals of X are incomparable, so are the prime ideals of X'. Thus X' is a right classical set.

In [LS1], it was actually shown that a right classical set of prime ideals of R transfer to a right classical set of prime ideals of T and then to a right classical set of prime ideals of S.

Again in [LS1], the above general results were applied to the fixed ring example of a Morita context (see Example 1.6 above). For instance, it was proved that if R is two-sided Noetherian and satisfies the second layer condition, then the two-sided Noetherian ring R^G also satisfies the second layer condition. Moreover, if $J(R)$ is classical, then so is $J(R^G)$.

We can improve on 3.5 if we make an additional assumption. First recall that a two-sided ideal I of a ring R has the *right AR-property* if for every right ideal K of R, there is a positive integer n such that $K \cap I^n \subseteq KI$ (see [GW] for more information). If I has the AR property on both sides, we will omit the mention of a side. This property is interesting because in a two-sided Noetherian ring, if I has the AR property, then whenever P is linked to Q and I is contained in one of these prime ideals, then I must actually be contained in both [Jt3, 5.3.10]. Thus we can improve on 3.5 if the ideal $(1 - e)TeT(1 - e)$ of S has the AR property.

3.8 Theorem. Let T be a two-sided Noetherian ring and assume that $I = (1 - e)TeT(1 - e)$ satisfies the AR property. Let X be a right link closed set of prime ideals of R satisfying the second layer condition. Then the corresponding set X' of prime ideals of S is also right link closed and satisfies the second layer condition.

Proof. Suppose that Q_1 is in X' and Q_2 is right linked to Q_1. It is clear from 3.5 that if Q_2 is not in X', then it must contain I. However, as previously indicated, this implies that Q_1 contains I, which is impossible since the prime ideals of S corresponding to prime ideals of R cannot contain I. Thus X' is right link closed. Finally X' satisfies the second layer condition as mentioned in the remark after 3.4.

An example where the hypotheses of 3.8 are satisfied can be found using subidealizers (Example 1.5). Let ideal of the two-sided Noetherian ring R and suppose that I satisfies the AR . If S is any two-sided Noetherian subring of R containing I, then it is not d. show that I, as an ideal of S, has the AR property. Thus the ring $T = \begin{pmatrix} R & R \\ I & S \end{pmatrix}$ the hypotheses of 3.8.

Finally, for an application of Morita contexts to the transfer of homological dimensions from R to S, with applications to fixed rings (Example 1.6 above) and subidealizers (Example 1.5 above), see [LS2].

REFERENCES

Am] S.A. Amitsur, "Rings of quotients and Morita contexts", J. Algebra **17** (1971) 273-298.

[An] F.W. Anderson, "Endomorphism rings of projective modules", Math. Z. III (1969) 322-332.

[B1] H. Bass, "The Morita theorems", mimeographed notes, University of Oregon, 1962.

[B2] H. Bass, Algebraic K-theory, W.A. Benjamin, New York, 1968.

[Be] A.D. Bell, "Notes on localization in Noetherian rings", Cuadernos de Algebra, Granada, Spain (1990).

[Br] K.A. Brown, On sets in Noetherian rings, Seminaire d'Algebre P. Dubreil et M.P. Malliavri 1983-1984, Springer-Verlag Lecture Notes in Mathematics **1146** (1985) 355-366.

[Co] M. Cohen, "A Morita context related to finite automorphism groups of rings", Pac. J. Math. **98** (1), (1982) 37-54.

RT] R.S. Cunningham, E.A. Rutter and D.R. Turnidge, "Rings of quotients of endomophism rings of projective modules", Pacific J. Math. **41** (1972) 647-668.

[Cu] R.S. Cunningham, "Morita equivalent rings of quotients", Notices Amer. Math. Soc. **19** (1972) 1-73.

[F] T.V. Forsum, "Lattice isomorphism between Morita related modules", Notices Amer. Math. Soc. **18** (1971) 361.

[Ga] E.M. Garcia-Herreros, Semitriviale Erweiterungen und Generalisierte Matrizenringe, Algebra-Berichte 54, Fischer (1986).

[Gd1] A.W. Goldie, "The structure of prime rings under ascending chain conditions", Proc. London Math. Soc. **8** (1958) 589-608.

[Gd2] A.W. Goldie, "Semi-prime rings with maximum condition", Proc. London Math. Soc. **10** (1960) 201-220.

[Gl] J.S. Golan, Torsion theories, Pitman Monographs in Pure and Applied Mathematics 29, New York (1986).

[Go] K.R. Goodearl, "Linked injectives and Ore localization", J. London Math. Soc. **37** (2), (1988) 404-420.

[Gu] R.N. Gupta, "On primitivity of matrix rings", Amer. Math. Monthly **75** (1968) 636.

[GW] K.R. Goodearl and R.B. Warfield, An introduction to noncommutative Noetherian rings, London Mathematical Society Student Texts 16, Cambridge University Press, Cambridge (1989).

[H] J.J. Hutchinson, "The completion of a Morita context", Comm. Algebra **8** (8), (1980) 717-742.

[HF] J.J. Hutchinson and M.H. Fenrick, "Primary decomposition and Morita contexts", Comm. Algebra **6** (13), (1978) 1359-1368.

[HL] J.J. Hutchinson and H.M. Leu, "Rings of quotients of R and eRe", Chinese J. Math. **4** (1), (1976) 25-35

[Hu] S.A. Hu, "Generalized matrix rings over a ring R and their application", J. Math (Wuhan) **7** (4), (1987) 407-415.

[HZ] J.J. Hutchinson and J. Zelmanowitz, "Quotient rings of endomorphism rings of modules with zero singular submodules", Proc. Amer. Math. Soc. **35** (1972) 16-20.

[Jc] N. Jacobson, "Structure of rings", Amer. Math. Soc. Colloq. Publ. **37** (1964).

[Jg1] M. Jagermann, "Morita contexts and radicals", Bull. Acad. Polon. Sci. **20** (1972) 619-625.

[Jg2] M. Jagermann, "Normal radicals of endomorphism rings of free and projective modules", Fund. Math. **86** (1975) 237-250.

[Jt1] A.V. Jategoankar, "Endomorphism rings of torsionless modules", Trans. Amer. Math. Soc. **161** (1971) 457-466.

[Jt2] A.V. Jategoankar, "Soluble Lie algebra, polycyclic-by-finite groups, and bimodule Krull dimension", Comm. Algebra **10** (1982) 19-69.

[Jt3] A.V. Jategoankar, Localization in Noetherian rings, London Mathematical Society, Lecture Notes 98, Cambridge University Press, Cambridge (1986).

[Ka1] A.I. Kashu, "Morita contexts and torsion modules", Math. Notes Acad. Sci. USSR **28** (1980) 706-710.

[Ka2] A.I. Kashu, "Jansian torsion and ideal torsion in Morita context", Mat. Issled. **62** (1981) 65-75.

[Ka3] A.I. Kashu, "On localizations in Morita contexts", Mat. Sb. (N.S.) **133** (175) (1) (1987) 127-133, 144.

[Kh1] S.M. Khuri, "Endomorphism rings and lattice isomorphism", J. Algebra **56** (1979) 401-408.

[Kh2] S.M. Khuri, "Endomorphism rings and Gabriel topologies", Can. J. Math, Vol. **XXVI**, No. 2 (1984) 193-205.

[KO] T. Kato and K. Ohtake, "Morita contexts and equivalences", J. Algebra **61** (1979) 360-366.

[Kt1] T. Kato, "Dominant modules", J. Algebra **14** (1970) 341-349.

[Kt2] T. Kato, "U-distinguished modules", J. Algebra **25** (1973) 15-24.

[Kt3] T. Kato, "Morita contexts and equivalences II", in *Proceedings of the 20th Symposium on Ring Theory*, Okayama University (1987) 31-36.

[LH] H.M. Leu and J.J. Hutchinson, "Kernel functors and quotient rings", Bull. Inst. Math. Acad. Sinica **5** (1), (1977) 145-155.

[LS1] P. Loustaunau and J. Shapiro, "Localization in a Morita context with applications to fixed rings", to appear in J. Algebra.

[LS2] P. Loustaunau and J. Shapiro, "Homological dimensions in a Morita context with applications to subidealizers and fixed rings", to appear in Proc. Amer. Math. Soc.

[cCR] J.C. McConnell and J.C. Robson, Noncommutative Noetherian rings, Wiley Series in Pure and Applied Mathematics, New York (1987).

[Mo] K. Morita, "Duality for modules and its applications to the theory of rings with minimum condition", Sci. Rep. Tokyo Kyoiku Diagaku Sect. **A6** (1958) 83-142.

[Mb] B.J. Müller, "The quotient category of a Morita context", J. Algebra **28** (1974) 389-407.

[Mm] M. Müller, "Rings of quotients of generalized matrix rings", Comm. Algebra **15** (10), (1987) 1991-2015.

[NW1] W.K. Nicholson and J.F. Watters, "Normal radicals and normal classes of rings", J. Algebra **59** (1979) 5-15.

[NW2] W.K. Nicholson and J.F. Watters, "Morita context functors", Math. Proc. Cambridge Philos. Soc. **103** (3), (1988) 399-408.

[Pa] I. Palmer, "The global homological dimension of semi-trivial extensions of rings", Math. Scand. **37** (1975) 223-256.

[Po] E.C. Posner, "Primitive matrix rings", Arch. Math. **12** (1961) 97-107.

[Sd] F.F. Sandomierski, "Modules over the endomorphism ring of a finitely generated projective module", Proc. Amer. Math. Soc. **31** (1972) 27-31.

[Sk1] K. Sakano, "Maximal quotient rings of generalized matrix rings", Comm. Algebra **12** (16), (1984) 1055-1065.

[Sk2] K. Sakano, "Some aspects of the ϕ-trivial extensions of rings", Comm. Algebra **13** (10), (1985) 2199-2210.

[Sk3] K. Sakano, "On cogenerator rings as ϕ-trivial extensions", Tsukuba J. Math. **11** (1), (1987) 121-130.

[Sn] A.D. Sands, "Radicals and Morita contexts", J. Algebra **24** (1973) 335-345.

[St] B. Stenström, "The maximal ring of quotients of a generalized matrix ring", Universale Algebren and Theorie der Radikale, pp. 65-67, Studien zur Algebra und ihre Anwendungen, Band 1, Akademie-Verlag, Berlin, 1976.

[Sw] P.N. Stewart, "Morita contexts and rings with unique maximal ideal", Math. Chronicle **16** (1987) 85-87.

[W] E.R. Willard, "Properties of projective generators", Math. Ann. **158** (1965) 352-364.

[Z] J.M. Zelmanowitz, "Endomorphism rings of torsionless modules", J. Algebra 5 (1967) 325-341.

George Mason University
Fairfax, VA 22030

ON THE WEAK RELATIVE-INJECTIVITY OF RINGS AND MODULES

Abdullah Al-Huzali, S.K. Jain and S.R. López-Permouth

1. INTRODUCTION

This paper is about two strongly related topics, right CEP-rings and weakly-injective rings and modules. Rings whose cyclic modules are embeddable essentially in direct summands and rings whose cyclics are embeddable essentially in projectives have been studied in [6], [7] and [8]. Following [6], we refer to the latter as right CEP-rings. We propose refering to the former as right CES-rings. In the above mentioned references, semiperfect CES-rings are characterized as follows:

Theorem A. Let R be a semiperfect ring. R is right CEP if and only if R is right artinian and every summand of R is weakly R-injective.

Theorem B. The following statements about a ring R are equivalent:

(i) R is semiperfect and right CES,

(ii) every ring homomorphic image of R is right CES,

(iii) every ring homomorphic image of R is right CEP,

(iv) R is of one of the following types:

   (a) R is uniserial as a right module,

   (b) R is an $n \times n$ matrix ring over a right self-injective ring of the type in (a), or

   (c) R is a direct sum of rings of types (a) or (b).

One is left to wonder about the necessity of the assumption that R is semiperfect. In this paper we show that right CES-rings and right CEP-rings are necessarily semiperfect in many special cases including when R is semiprime, right nonsingular, right semi-hereditary, right self-injective, or one-sided Noetherian.

The concept of weak relative-injectivity of modules as mentioned in Theorem A was first introduced in [7]. Its study has been furthered in [8] and [5]. The second reference deals with the somewhat more general concept of tight modules. Weak relative-injectivity of modules is closed under finite direct sums but it notably fails to be inherited by direct summands. The same holds true for tightness. We provide examples of a countably infinite collection of modules none of which is weakly-injective whose sum is weakly-injective. Also, we investigate when the condition of weak-injectivity on (the summands of) a semiperfect ring is equivalent to the injectivity of the ring. Our results include the fact that a semiperfect ring R is right self-injective if and only if each summand of R is weakly-injective and the Jacobson radical of R coincides with its right singular ideal. Also, we show that a right or left perfect ring is self-injective if and only if it is weakly-injective.

Throughout this paper R is a ring with 1 and all modules are right and unital unless otherwise specified. As usual, $J(M)$, $Z(M)$ and $E(M)$ denote respectively the Jacobson radical, the singular submodule and the injective hull of a module M. Any term not defined here may be found in a standard reference such as [1].

2. CEP-RINGS

Given two modules M and N, we say that M is weakly N-injective if for every map $f : N \to E(M)$, one may find a submodule X of $E(M)$ which is isomorphic to M and satisfies that the image of f lies in X. We also say that M is N-tight if every homomorphic image of N which is embeddable in $E(M)$ is also embeddable in M. Obviously, every weakly N-injective module is N-tight as well. If a module is weakly R^n-injective (R^n-tight) for all positive integer n, we say that M is weakly-injective (tight). When M is cyclic one easily sees that M is weakly-injective (tight) if and only if it is weakly R^2-injective (R^2-tight).

The exact relationship between weak-injectivity and tightness is given by the following lemma from [8].

2.1 Lemma. Given two right modules M and N, M is weakly N-injective if and only if for every submodule Q of N and for every monomorphism $\sigma : N/Q \to E(M)$

(i) there exists a monomorphism $\sigma' : N/Q \to M$ and

(ii) for every complement K of $\sigma'(N/Q)$ in M, there exists a submodule K' of $E(M)$ such that $K' \cap \sigma(N/Q) = 0$ and $K' \cong K$.

Proof. See [8]. □

2.2 Corollary. A uniform module M is weakly N-injective if and only if it is N-tight.

Proof. Obvious. □

In view of this corollary, the following characterization of right CEP-rings follows from Proposition 1.12 in [8].

2.3 Proposition. A semiperfect ring R is right CEP if and only if R is right artinian and every projective indecomposable right R module is uniform and tight.

Proof. Immediate from the discussion. □

If one assume that every right module is tight, then, in particular, every uniform right module is weakly-injective. Also, under this hypothesis, completely reducible right R-modules would have to be injective which implies that R is right Noetherian. In light of Theorem 2.5 in [8], we arrive to the following proposition.

2.4 Proposition. A ring R is right weakly-semisimple (i.e., every right modules is weakly-injective) if and only if any one of the following equivalent conditions holds:

(i) every right R-module is tight;

(ii) R is right Noetherian and every finitely generated right R-module is tight;

(iii) R is right Noetherian and every cyclic right R-module is tight; or

(iv) R is right Noetherian and every uniform cyclic right R-module is tight.

Proof. Every right weakly-semisimple ring satisfies (i). By the above discussion, (i) implies (ii). Clearly, (ii) implies (iii) and (iii) implies (iv). By Corollary 2.2, (iv) is equivalent to "R is right Noetherian and every uniform cyclic right R-module is weakly R-injective "; hence by Theorem 2.5 in [8], (iv) implies that R is weakly-semisimple. □

The following theorems show that rings whose cyclics are embeddable in free modules (thus right CEP-rings) are semiperfect under any of several additional hypotheses. Some of these results are implicit in [2].

2.5 Theorem. A ring R for which every cyclic right module embeds in a free is semisimple artinian if R is semiprime, right nonsingular or right semi-hereditary.

Proof. Suppose R is not semisimple artinian and let M be a maximal essential right ideal. By hypothesis, R/M embeds in a free module F via a monomorphism $\varphi : R/M \to F$. Easily one concludes that F must be a finite sum $F = R \oplus \cdots \oplus R$. Since R is simple one of the projection maps $\pi : F \to R$ satisfies that $\pi \varphi(R/M) \cong R/M$. In the case when R is semiprime, $\varphi(R/M) = eR$, where $e^2 = e$ is an idempotent. This implies that $(1 - e)R$, the right annihilator of e, equals M. This contradicts the assumption that M is essential. We conclude that when R is semiperfect R is semisimple-artinian. Consider now the case when R is right nonsingular. In this case, R/M is isomorphic to a single right idea aR of R, with $r(a) = M$. Since M is assumed to be semisimple, one concludes that a belongs to the right singular ideal $Z(R)$. This contradicts the nonsingularity of R. Once again, we conclude that R must be semisimple artinian. Lastly, suppose R is right semi-hereditary (*i.e.*, every principal right ideal is projective). Since R/M is isomorphic to a simple right ideal aR or R and aR is projective, one concludes that M splits in R, contradicting its essentiality. Therefore, if R is semi-hereditary, it must be semisimple artinian. □

2.6 Theorem. A ring R for which every cyclic right module embeds in a free is artinian whenever R is right self-injective or one-sides Noetherian.

Proof. If R is right self-injective the result follows Theorem 3.3 in [7], since every right self-injective ring is right QF-3. Alternatively, the result follows from the fact that a right self-injective ring satisfying the hypothesis is right PF and hence semiperfect [11]. When R is left Noetherian it satisfies the ascending chain condition (ACC) on annihilator left ideals and, equivalently, the descending chain condition (DCC) on annihilator right ideals. The result follows since, under our hypothesis, every right ideal is an annihilator right ideal. Finally, assume that R is right Noetherian. By Lemma 4 in [10], R^n has essential right socle for every positive integer n. Since every cyclic embeds in some R^n we conclude that every right module has nonzero socle. It follows that R is left perfect and hence, by Hopkin's theorem, right artinian. □

A right Noetherian ring for which every cyclic right module is an annihilator does not need to be right artinian [4].

3. WEAK-INJECTIVITY

As mentioned in the introduction, any finite direct sum of weakly-injective (tight) modules is weakly-injective (tight) but a direct summand of a weakly-injective (tight) module may not be weakly-injective (tight). Indeed, any module M over an arbitrary ring is a direct summand of a tight module $T = M \oplus H$, if we take H to be the direct product of an infinite number of copies of the injective hull of M. Similarly, any module M over a right Noetherian ring R is a summand of a weakly-injective module $W = M \oplus S$. Simply take S to be a direct sum of infinitely many copies of the injective hull of M. In order to illustrate further how weak-injectivity is not transferred down to summands, we present next an example of a countable family of non-weakly-injective modules whose sum is weakly-injective.

3.1 Example. Let $M_i = \mathbf{Z}/2^i \mathbf{Z}$. The (external) direct sum $M = \oplus \sum_{i=1}^{\infty} M_i$ is a weakly-injective but for all i, M_i is not weakly-injective.

Proof. Since M_i is finite while its injective hull $E(M_i) = \mathbf{Z}(2^\infty)$ is not, obviously M_i is not even tight. As usual, we shall identify $\mathbf{Z}/2^i \mathbf{Z}$ with the submodule $\{0, \frac{1}{2^i}, \frac{2}{2^i}, \cdots$ $, \frac{2^{i-1}}{2^i} \}$ of $\mathbf{Z}(2^\infty)$. Then, the $M_i's$ constitute an ascending chain $M_1 \subseteq M_2 \subseteq \cdots$ whose union is $\mathbf{Z}(2^\infty)$. Let $x_1, \cdots, x_n \in E(M)$. Since $E(M)$ is a countably infinite sum $\oplus \sum_{i=1}^{\infty} \mathbf{Z}(2^\infty)$ of copies of $\mathbf{Z}(2^\infty)$, $x_1, \cdots, x_n$ belong to a finite sub-sum $N = \oplus \sum_{i=1}^{t} \mathbf{Z}(2^\infty)$. Since each copy of $\mathbf{Z}(2^\infty)$ is a union of the chain $M_1 \subseteq M_2 \subseteq \cdots$, one can choose inside the ith copy of $\mathbf{Z}(2^\infty)$ a submodule isomorphic to M_{n_i} in such a way that $M_{n_i} \not\cong M_{n_j}$ when $i \neq j$ and $x_1, \cdots x_n \in M_{n_1} \oplus \cdots \oplus M_{n_t}$. Out of the remaining summands of M pick copies of the $M_i's$ whose isomorphism classes are not represented in $\{M_{n_1}, \cdots, M_{n_t}\}$. the sum of these submodules plus the $M_{n_i}'s$ is a submodule X of $E(M)$ which is isomorphic to M and satisfies that $x_1, \cdots, x_n \in X$, proving that M is weakly-injective. $\square$

It is known that, for many rings, weak-injectivity is quite a distinct property from injectivity. For example, in [8] it has been shown that a domain R is weakly-injective if and only if it is two-sided Ore. However, in that same reference we see that if a ring R is right artinian, then R is right weakly-injective if and only if it is right self-injective (i.e., R is quasi-Frobenius). Our next theorems are concerned with then condition of weak-injectivity on the summands of a semi-perfect ring R is equivalent to the self-injectivity of R. For the remainder of the paper the semi-perfect ring R may be written as $R = \oplus \sum_{i=1}^{n} e_i R$ where End $(e_i R)$ is local for $i = 1, \cdots, n$.

3.2 Theorem. Let R be a semiperfect ring then

(i) R is self-injective if every summand of R is weakly-injective and R has nil Jacobson radical $J(R)$, and

(ii) R is self-injective if and only if every summand of R is weakly-injective and the Jacobson radical $J(R)$ coincides with the right singular ideal $Z_r(R)$.

Proof.

(i) Consider a projective indecomposable module eR with $e^2 = e$. By its weak-injectivity, we know that since e, $x \in E(eR)$ there exists $X \cong eR$ such that e, $x \in X$. This means that X is local, hence if $eR \neq X$ we must have that e belongs to the Jacobson radical $J(X)$ of X. So, there exists an embedding $\varphi : eR \to eJ$. Let $\varphi(e) = ea = a$. Let $n \in Z$ be the smallest such that $a^n = 0$. Then $\varphi(ea^{n-1}) = a \cdot a^{n-1} = 0$, but since φ is one-to-one $ea^{n-1} = a^{n-1} = 0$, a contradiction. We, therefore, conclude that $eR = X$ and hence $x \in eR$. Since x was chosen arbitrarily, we conclude that $eR = E(eR)$ is indeed injective.

(ii) It is well known that a self-injective ring satisfies that $J(R) = Z_r(R)$ and obviously each summand of R must be (weakly-) injective. Conversely, consider an indecomposable projective module eR, with $e^2 = e$. Arguing as before, for arbitrary $x \in E(eR)$, e and x belong to a local submodule X of $E(eR)$ which is isomorphic to eR. If $eR \neq X$ we get an embedding φ of eR into eJ. This leads to a contradiction for if $\varphi(e) = a \in J(R) = Z_r(R)$, then the right annihilator of a, $r(a)$, must be essential. However, $r(a) = r(e) = (1-e)R$. so $1-e$ is an element of an essential right ideal of R. This leads to a contradiction. Therefore, we conclude that $x \in eR$ and then $eR = E(eR)$, as desired. □

The following theorem concerns the weak-injectivity of a right or left perfect ring.

3.3 Theorem. A right or left perfect ring R is self-injective if and only if it is weakly-injective.

Proof. Assume that R is left perfect. This is equivalent to saying that R satisfies the DCC on principal right ideals. Let 1, $x \in E(R)$, and let X be a submodule of $E(R)$, isomorphic to R such that 1, $x \in X$. If $R \neq X$ then R is embeddable as a proper principal right ideal of itself. This would yield an infinite decreasing sequence of principal right ideals, a contradiction. Hence $R = X$ and, therefore, since x was picked arbitrarily one concludes that $R = E(R)$ is self-injective. On the other hand, assume that R is right perfect. A result of Jonah [9] states that this is equivalent to every right module of R satisfying the ACC on cyclic submodules. In particular, $E(R)$ satisfies such ACC.

As before let $x \in E(R)$ and let X be a submodule of $E(R)$, isomorphic to R such that 1 and x belong to X. If $R \neq X$ then X is a cyclic submodule of $E(R)$ which contains R properly. Yet, X is isomorphic to R via an isomorphism $\varphi : R \to X$, say. Extend φ to an isomorphism $\hat{\varphi} : E(R) \to E(X) = E(R)$. Then $\hat{\varphi}(X)$ is a cyclic submodule of $E(R)$ containing X properly. Continuing this way one gets an infinite ascending chain of cyclic submodules of $E(R)$, a contradiction. Thus, $R = X$ and since x was chosen arbitrarily $R = E(R)$ is self-injective. □

3.4 Corollary. A semiprimary ring R is self-injective if and only if it is weakly-injective.

Proof. Semiprimary rings are right and left perfect.

3.5 Corollary. A right or left artinian ring R is quasi-Frobenius if and only if it is weakly-injective.

Proof. Right or left artinian rings are semiprimary.

Acknowledgement. The third author wishes to thank partial support from the Ohio University Research Committee under grant # 848.

REFERENCES

1. F.W. Anderson and K.R. Fuller, *Rings and categories of modules*, Springer-Verlag, New *York/Heidelberg/Berlin*, 1974.
2. C. Faith, Embedding modules in projectives: A report on a problem", in *Advances in Non-Commutative Ring Theory*, Vol. 951, Lecture Notes in mathematics, Springer-Verlag, New *York/Berlin/Heidelberg*, 1981.
3. C. Faith, "When self-injective rings are QF: A report on a problem", preprint.
4. C. Faith and P. Menal, "Counter-example to a conjecture of John's", preprint.
5. J.S. Golan and S.R. López-Permouth, "A remark on QI-filters and tight modules", to appear.
6. S.K. Jain and S.R. López-Permouth, "A generalization of the Wedderburn-Artin theorem", Proc. Am. Math. Soc. **106** (1) (1989) 19-23.
7. S.K. Jain and S.R. López-Permouth, "Rings whose cyclics are essentially embeddable in projectives modules" , J. of Algebra **128** (1) (1990) 257-269.
8. S.K. Jain, S.R. López-Permouth and Surjeet Singh, "On a class of QI-rings", to appear.
9. D. Jonah, "Rings with minimum condition for principal right ideals have the maximum condition for principal left ideals", Math. Z. **113** (1970) 106-112.
10. B. Johns, "Annihilator conditions in Noetherian rings", J. Algebra **49** (1977) 222-224.
11. B.L. Osofsky, "A generalization of Quasi-Frobenius rings", J. Algebra **4** (1966) 373-387; Erratum **9** (1968), 120.

CS-MODULES AND WEAK CS-MODULES

Patrick F. Smith

Let R be a ring and M a right R-module. The module M is a CS-module if every submodule is essential in a direct summand. We shall call the module M a weak CS-module provided every semisimple submodule is essential in a direct summand. We prove that the direct sum of a semisimple module and an injective module is a weak CS-module, but such a module is not, in general, a CS-module, even for Dedekind domains. Moreover, for any Dedekind domain R, every module with finite uniform dimension is a weak CS-module.

Weak CS-modules share some of the properties of CS-modules. For example, we prove that if M is a weak CS-module which satisfies the ascending chain condition on essential submodules then M is a direct sum of a semisimple module and a Noetherian module.

We also prove that if R is a ring with the property that every cyclic right R-module is a direct sum of a projective module, a CS-module and a module with finite uniform dimension then every cyclic right R-module has finite uniform dimension. Similarly, if R is a ring such that every cyclic right R-module is a direct sum of a projective module, an injective module and a module with Krull dimension, then the ring R has right Krull dimension.

1. WEAK CS-MODULES

In what follows, all rings R have identities and all modules are unital right R-modules. Unless otherwise stated, the rings considered are arbitrary.

Let R be a ring. Let M be an R-module. Let N be a submodule of M. By Zorn's Lemma there exists a submodule K of M maximal with respect to the property $K \cap N = 0$, and we call K *a complement of N in M*. Note that, in this case, $N \oplus K$ is an essential submodule of M[1, Proposition 5.21]. A submodule K of M is called a *complement in M* (or a *complement submodule of M*) if there exists a submodule N of M such that K is a complement of N in M. It is not difficult to prove that K is a complement in M if and only if K has no proper essential extension in M. For example, any direct summand of M is a complement.

The module M is called a *CS-module* if every complement is a direct summand of M; equivalently, every submodule is essential in a direct summand of M. Among examples of CS-modules, we could mention semisimple modules, uniform modules and quasi-injective modules. Note that CS-modules are termed "extending modules" in [6], etc. and "modules with property (C1)" in [15]. CS-modules have been studied by various authors; apart from the references already cited, see [2], [4], [7], [9], [11], [12], [13], [16] and [17] (for good sources of references, see [15] or [20]).

An R-module M will be called a *weak CS-module* provided, for each semisimple submodule S of M, there exists a direct summand K of M such that S is essential in K. There is another class of modules which we shall consider in this section, namely the class of modules M with the following property:

Every complement with essential socle is a direct summand, and we shall call these modules *CESS-modules*. The following implications are clear, for any module M :

M is a CS-module $\Rightarrow$ M is a CESS-module $\Rightarrow$ M is a weak CS-module.

Any module with zero socle is a CESS-module. Thus, for the ring $\mathbf{Z}$ of rational integers, any free $\mathbf{Z}$-module of infinite rank is a CESS-module which is not a CS-module (see [11, Theorem 5]). Note further that any CESS-module with essential socle is a CS-module. Next we shall give, also for the ring $\mathbf{Z}$, an example of a weak CS-module which is not a CESS-module.

Example 1.1. Let p be any rational prime and M the $\mathbf{Z}$-module $(\mathbf{Z}/\mathbf{Z}p) \oplus (\mathbf{Z}/\mathbf{Z}p^3)$. Then M is a weak CS-module which is not a CESS-module. Note further that every proper submodule of M is a CS-module.

Proof. Note that M has essential socle. By [6, Theorem 7] or [12, Corollary 23], M is not a CS-module and hence not a CESS-module. (In fact, the submodule $K = \mathbf{Z}(1+\mathbf{Z}p, p + \mathbf{Z}p^3)$ is a complement submodule of M of order p^2. If K were a direct summand of M, then $M = K \oplus K'$, for some submodule K' of M, and hence K' has order p^2 also, giving $p^2 M = 0$, a contradiction. Thus M is not a CS-module.)

Next, we shall prove that M is a weak CS-module. Note that M has uniform dimension 2. Let S be a semisimple submodule of M. If S is not simple, then S is essential in M itself. Suppose that S is simple. Then $S = \mathbf{Z}(a + \mathbf{Z}p, p^2 b + \mathbf{Z}p^3)$, for some integers a, b such that $0 \leq a, b \leq p - 1$. If $a = 0$, then S is essential in the direct summand $L = 0 \oplus (\mathbf{Z}/\mathbf{Z}p^3)$ of M. If $a \neq 0$, then $M = S \oplus L$. Thus, in any case, S is essential in a direct summand of M. It follows that M is a weak CS-module.

Now, let N be any proper submodule of M. Then N has order at most p^3, so that N is a CS-module by [6], Theorem 7.

Next we shall prove, for the sake of completeness, the following basic fact (see, for example, [4, Proposition 2.2]).

Lemma 1.2. Let R be any ring and $L \subseteq K$ be submodules of a right R-module M. Suppose further that L is a complement submodule of K and K a complement submodule of M. Then L is a complement submodule of M.

Proof. There exist submodules K' of M and L' of K such that K is a complement of K' in M and L is a complement of L' in K. Note that

$$L \cap (K' + L') = 0 .$$

Suppose that N is a submodule of M such that $L \subseteq N$ and $N \cap (K' + L') = 0$. Then $(N + K') \cap L' = 0$, and hence $(N + K') \cap K = L$. It follows that $(N + K) \cap K' = 0$, and hence $K = N + K$. Now $N \subseteq K$ and $N \cap L' = 0$ together imply $N = L$. It follows that L is a complement of $K' + L'$ in M.

Corollary 1.3. Any direct summand of a CS-module (respectively, CESS-module) is also a CS-module (respectively, CESS-module).

Proof. By Lemma 1.2.

We have been unable to settle the following:

Question 1.4. Let K be a direct summand of a weak CS-module M. Is K a weak CS-module?

Our next result gives some information about CESS-modules. For any module M, Soc M will denote the socle of M.

Proposition 1.5. Let M be a module with socle S. Then M is a CESS-module if and only if every complement K in M, with S essential in K, is a CS-module and a direct summand of M.

Proof. Suppose first that M is a CESS-module. Let K be a complement in M with S essential in K. Then K is a direct summand of M. Moreover, K has essential socle and, by Corollary 1.3, K is a CESS-module. Thus K is a CS-module.

Conversely, suppose that M has the stated property. Let N be any complement submodule of M with essential socle. There exists a submodule L of M such that $S = (Soc\ N) \oplus L$. Note that S is essential in $N \oplus L$. There exists a complement K in M such that $N \oplus L$ is essential in K. Thus S is essential in K, and, by hypothesis, K is a CS-module and is a direct summand of M. But N is a complement in K, and hence N is a direct summand of K. Thus N is a direct summand of M. It follows that M is a CESS-module.

Corollary 1.6. Let M be a CESS-module. Then $M = M_1 \oplus M_2$ for some CS-module M_1 with essential socle and module M_2 with zero socle.

Proof. Let $S = Soc\ M$. There exists a complement M_1 in M such that S is essential in M_1. Then M_1 is a direct summand of M and a CS-module by Proposition 1.5. The result follows.

Now we ask:

Question 1.7. Is the converse of Corollary 1.6 true?

We next show that the converse of Corollary 1.6 is true if we restrict our attention to modules over (commutative) Dedekind domains.

Proposition 1.8. Let R be a Dedekind domain. Then an R-module M is a CESS-module if and only if $M = M_1 \oplus M_2$ for some torsion CS-module M_1 and torsion-free module M_2.

Proof. Note that, over the ring R, a module has essential socle if and oly if it is a torsion module. Thus the necessity follows by Corollary 1.6. Conversely, suppose that

$M = M_1 \oplus M_2$ for some torsion CS-module M_1 and torsion-free module M_2. Let L be a complement in M with essential socle. Then $L \subseteq M_1$. But, clearly, L is a complement in M_1. It follows that L is a direct summand of M_1, and hence also of M. Thus M is a CESS-module.

For Dedekind domains, CS-modules are classified (see [6, Theorem 7], [11, Theorem 14], and [12, Corollary 23]). Let R be a Dedeking domain and M an R-module. For any maximal ideal P of R let

$$M(P) = \{m \in M : P^n m = 0 \text{ for some positive integer } n\} .$$

Proposition 1.9. Let R be a Dedekind domain. Then an R-module M is a CS-module if and only if either

(i) M is a torsion module such that, for each maximal ideal P of R, $M(P)$ is injective or $M(P)$ is a direct sum of copies of R/P^n or R/P^{n+1}, for some positive interger $n = n(P)$, or

(ii) M is not torsion, and $M = M_1 \oplus NI_1 \oplus \cdots \oplus NI_k$, where M_1 is an injective module, k a positive integer, N a proper R-submodule of the quotient field K of R and $I_1, \cdots, I_k$ fractional ideals of R.

It is clear from Propositions 1.8 and 1.9 that if R is a Dedeking domain and M an R-module such that M is a CS-module (respectively, a CESS-module) and M' a semisimple or injective module, then $M \oplus M'$ need not be a CS-module (respectively, a CESS-module). Now we return to the consideration of modules over general rings. The next two results show that weak CS-modules are better behaved.

Lemma 1.10. Let M be a module such that $M = M_1 \oplus M_2$ for some weak CS-module M_1 and semisimple module M_2. Then M is a weak CS-module.

Proof. Let S be any semisimple submodule of M. Note first that

$$S + M_1 = M_1 \oplus [(S + M_1) \cap M_2] .$$

Because M_2 is semisimple, the module $(S + M_1) \cap M_2$ is a direct summand of M_2 ([1, Theorem 9.6]). Thus $S + M_1$ is a direct summand of M.

Now $S \cap M_1$ is a submodule of S, so that, again applying [1, Theorem 9.6], there exists a submodule S' of S such that $S = (S \cap M_1) \oplus S'$. Since M_1 is a weak CS-module, it follows that there exists a direct summand K of M_1 such that $S \cap M_1$ is essential in K. But

$$S + M_1 = (S \cap M_1) + S' + M_1 = M_1 \oplus S' .$$

Thus $S = (S \cap M_1) \oplus S'$ is an essential submodule of the direct summand $K \oplus S'$ of $S + M_1$. It follows that S is essential in a direct summand of M. Therefore, M is a weak CS-module.

Lemma 1.11. Let M be a module such that $M = M_1 \oplus M_2$ for some weak CS-module M_1 and injective module M_2. Then M is a weak CS-module.

Proof. Let S be any semisimple submodule of M. By [1, Theorem 9.6], there exists a submodule S' of S such that $S = (S \cap M_2) \oplus S'$. Note that $S' \cap M_2 = 0$, and hence $(M_2 + S')/S' \cong M_2$, an injective module. Thus there exists a submodule M' of M, containing S', such that

$$M/S' = [(M_2 + S')/S'] \oplus (M'/S') .$$

Note that $M = M_2 \oplus M'$. Thus $M' \cong M/M_2 \cong M_1$. It follows that M' is a weak CS-module.

There exists a direct summand K' of M' such that S' is essential in K'. Now $S \cap M_2$ is a submodule of the injective module M_2. Therefore, there exists a direct summand K of M_2 such that $S \cap M_2$ is essential in K (take K to be the injective envelope of $S \cap M_2$ in M_2). Now $S = (S \cap M_2) \oplus S'$ is essential in the direct summand $K \oplus K'$ of $M = M_2 \oplus M'$. It follows that M is a weak CS-module.

Theorem 1.12. Let $M = M_1 \oplus M_2 \oplus M_3 \oplus M_4$, where M_1 is a weak CS-module, M_2 a semisimple module, M_3 an injective module and M_4 a module with zero socle. Then M is a weak CS-module.

Proof. By Lemmas 1.10, 1.11 and [1, Proposition 9.19.]

Theorem 1.12 raises the following question:

Question 1.13. Let M be a module such that $M = M_1 \oplus M_2$, where M_1 and M_2 are both weak CS-modules. Is M a weak CS-module?

Note that, in particular, Theorem 1.12 shows that any direct sum of a semisimple module and an injective module is a weak CS-module. We know that such a module need not be a CS-module, in general, by Proposition 1.9. Perhaps it is worth highlighting this fact by giving a specific example.

Example 1.14. Let p by any rational prime, $M_1 = \mathbf{Z}/\mathbf{Z}p$ and $M_2 = \mathbf{Z}(p^\infty)$, the Prüfer p-group. Then the $\mathbf{Z}$-module $M = M_1 \oplus M_2$ is a weak CS-module which is not a CESS-module.

Proof. We shall suppose that $M_1 = \mathbf{Z}a$ and

$$M_2 = \mathbf{Z}b_1 + \mathbf{Z}b_2 + \mathbf{Z}b_3 + \cdots ,$$

where $pb_1 = 0$ and $pb_{i+1} = b_i (i \geq 1)$. Let $K = \mathbf{Z}(a, b_2)$. Then K is a complement in M. Suppose that $M = K \oplus K'$, for some submodule K' of M. Then

$$(0, b_3) = n(a, b_2) + x ,$$

for some $n \in \mathbf{Z}$ and $x \in K'$. It follows that

$$(0, b_1) = p^2(0, b_3) = p^2 x \in K' .$$

But $(0, b_1) = p(a, b_2) \in K$. Thus $K \cap K' \neq 0$, a contradiction. Thus M is not a CS-module. Since M has essential socle, it follows that M is not a CESS-module. However, by Theorem 1.12, M is a weak CS-module.

Now we return to the case of general Dedekind domains.

Lemma 1.15. Let R be a Dedekind domain, P a maximal ideal of R and M a non-zero finitely generated R-module such that $P^n M = 0$ for some positive integer n. Let S be a simple submodule of M. Then there exists a direct summand K of M such that S is an essential submodule of K.

Proof. It is known that M is a finite direct sum of cyclic submodules, say $M = Rm_1 \oplus \cdots \oplus Rm_k$, for some positive integer k and elements $m_i \in M (1 \leq i \leq k)$. For each $1 \leq i \leq k$, there exists $n(i) \geq 1$ such that $p^{n(i)} = \{r \in R : rm_i = 0\}$. Without loss of generality, we can suppose that

$$n(1) \leq n(2) \leq \cdots \leq nk = n .$$

Suppose that $n(1) = n$. Then M is a CS-module by Proposition 1.9, and thus S is essential in a direct summand of M.

Suppose that $n(1) \neq n$. There exists elements $a_i \in R (1 \leq i \leq k)$ such that $S = R(a_1 m_1 + \cdots + a_k m_k)$. Suppose that $a_1 m_1 = 0$. Then $S \subseteq Rm_2 \oplus \cdots \oplus Rm_k$, and, by induction on k, there exists a direct summand K of $Rm_2 \oplus \cdots \oplus Rm_k$, and hence also of M, such that S is essential in K. Thus we suppose $a_1 m_1 \neq 0$. Similarly, $a_i m_i \neq 0 (2 \leq i \leq k)$.

Note that $PS = 0$. In particular, for each $1 \leq i \leq k$, $Pa_i m_i = 0$, hence $Pa_i \subseteq p^{n(i)}$ and $a_i \in P^{n(i)-1}$, because P is invertible. There exists an ideal A_i of R such that $Ra_i = A_i P^{n(i)-1}$ and $R = A_i + P$, for each $1 \leq i \leq k$. It follows that

$$a_i M = P^{n(i)-1} (1 \leq i \leq k) ,$$

and hence

$$a_1 M \supseteq a_2 M \supseteq \cdots \supseteq a_k M .$$

In particular, for any $2 \leq i \leq k$,

$$a_i m_i = a_1 (b_{i1} m_1 + \cdots + b_{ik} m_k) ,$$

for some $b_{ij} \in R (1 \leq j \leq k)$. Thus $a_i m_i = a_1 b_{ii} m_i (2 \leq i \leq k)$.

Now

$$S = R(a_1 m_1 + \cdots + a_k m_k) = R(a_1 m_1 + a_1 b_{22} m_2 + \cdots + a_1 b_{kk} m_k)$$

$$= a_1 R(m_1 + b_{22}m_2 + \cdots + b_{kk}m_k) \ .$$

Let $K = R(m_1 + b_{22}m_2 + \cdots + b_{kk}m_k)$. Note that

$$M = K \oplus M_2 \oplus \cdots \oplus M_k \ .$$

Moreover, $S \subseteq K$, K is cyclic and $p^n K = 0$, so that P is uniserial and hence S is essential in K.

Theorem 1.16. Let R be a Dedekind domain. Then any finitely generated R-module is a weak CS-module.

Proof. Let M be a finitely generated R-module. Then $M = M_1 \oplus M_2$, for some torsion module M_1 and torsion-free module M_2. Let S be any semisimple submodule of M. Then $S \subseteq M_1$. Note that M_1 has finite (composition) length. Let U be a simple submodule of S. There exists a maximal ideal P of R such that $PU = 0$. Let $M' = \{m \in M : P^k m = 0 \text{ for some positive integer } k\}$. Then M' is a direct summand of M_1, and $P^n M' = 0$ for some positive integer n. Note that $U \subseteq M'$, so that, by Lemma 1.15, there exists a direct summand K of M' such that U is essential in K.

Now $M_1 = K \oplus K'$ for some submodule K' of M_1. Also, using [1, Proposition 9.19], we have:

$$S \subseteq Soc\ M_1 = (Soc\ K) \oplus (Soc\ K') = U \oplus (Soc\ K') \subseteq U \oplus K' \ .$$

Hence $S = U \oplus (S \cap K')$. But $S \cap K'$ is a semisimple submodule of K' and K' has smaller length than M_1. Thus, by induction of the length of M_1, there exists a direct summand L of K' such that $S \cap K'$ is essential in L. Then $S = U \oplus (S \cap K')$ is an essential submodule of the direct summand $K \oplus L$ of M. It follows that M is a weak CS-module.

Corollary 1.17. Let R be a Dedekind domain and M an R-module with finite uniform dimension. Then M is a weak CS-module.

Proof. If M is torsion-free, then M has zero socle and, in this case, M is certainly a weak CS-module. Now suppose that M is not torsion-free. By [14, Theorem 9], it follows that the module $M = M_1 \oplus M_2$, for some finitely generated module M_1 and injective submodule M_2. By Theorem 1.16 and Lemma 1.11, M is a weak CS-module.

2. EVENTUALLY SEMISIMPLE MODULES

Let R be a ring and M an R-module. Following [2], we call M *eventually semisimple* provided, for every direct sum

$$M_1 \oplus M_2 \oplus M_3 \oplus \cdots$$

of submodules $M_i (i \geq 1)$ of M, there exists a positive integer k such that M_i is semisimple for all $i \geq k$. Clearly semisimple modules and modules with finite uniform dimension are examples of eventually semisimple modules. Moreover, any eventually semisimple module with zero socle has finite uniform dimension. Camillo and Yousef [2, Lemma 1] prove that if $M/(Soc \, M)$ has finite uniform dimension then M is eventually semisimple. The converse is false (see, for example, [10, Example 6.3.15]). Other examples of eventually semisimple modules are given by the following result.

Lemma 2.1. Let $M = M_1 \oplus M_2$ where M_1 is a semisimple module and M_2 a module with finite uniform dimension. Then the module M is eventually semisimple.

Proof. Let $N_1 \oplus N_2 \oplus N_3 \oplus \cdots$ be a direct sum of submodules of M. For each $1 \leq i \leq j$, let

$$K_i = N_i \oplus N_{i+1} \oplus \cdots, \quad \text{and} \quad K_{ij} = N_i \oplus \cdots \oplus N_j \, .$$

Suppose that $K_1 \cap M_2 \neq 0$. Then $K_{1S} \cap M_2 \neq 0$ for some $S \geq 1$. Suppose that $K_{S+1} \cap M_2 \neq 0$. Then $K_{S+1,t} \cap M_2 \neq 0$ for some $t \geq S + 1$. Repeating this argument we produce a direct sum

$$K_{1S} \oplus K_{S+1,t} \oplus \cdots$$

of non-zero submodules of M_2. Since M_2 has finite uniform dimension, it follows that this process must stop. Thus there exists a positive integer q such that $K_q \cap M_2 = 0$. Now K_q embeds in M_1, and thus K_q is semisimple. It follows that M is eventually semisimple.

We shall call a module M *almost semisimple* if

(i) M has essential socle and

(ii) every finitely generated semisimple submodule of M is a complement in M.

Clearly semisimple modules are almost semisimple. On the other hand, let R be a commutative von Neumann regular ring which is not Noetherian. There exists a semisimple module S which is not injective [1, Proposition 18.13]. Let M denote the injective hull of S. Note that M is not semisimple but has essential socle. Moreover, M is almost semisimple because every finitely generated semisimple submodule of M is injective and hence a direct summand of M, by [1, p. 216 ex. 23].

On the other hand if R is a right Noetherian ring, then every almost semisimple right R-module is semisimple, as the following result shows. A module is called *locally Noetherian* if every finitely generated submodule is Noetherian.

Lemma 2.2. Any locally Noetherian almost semisimple module is semisimple.

Proof. Let M be a locally Noetherian almost semisimple module. Let $m \in M$. Then mR is Noetherian. Thus Soc mR is finitely generated and thus is a complement in M, by hypothesis. It follows that Soc mR is a complement in mR. But Soc mR is essential in mR. Thus $mR = Soc\ mR$. It follows that M is semisimple.

Lemma 2.3. Let M be an eventually semisimple module. Then there exists an almost semisimple complement K in M such that M/K has finite uniform dimension.

Proof. Let $S = Soc\ M$. If $S = 0$ then M has finite uniform dimension so we take $K = 0$. Suppose that $S \neq 0$. Let N_1 be a complement of S in M. Because $N_1 \cap S = 0$ it follows that N_1 has finite uniform dimension. Let K_1 be a complement of N_1 in M with $S \subseteq K_1$. Note that, by [1, Proposition 5.20], S is essential in K_1. Suppose that K_1 is not almost semisimple. Then there exists a finitely generated submodule S_1 of K_1 such that S_1 is not a complement in K_1. Let N_2 be a maximal essential estension of S_1 in K_1. Then $S_1 \neq N_2$, $N_1 \cap N_2 = 0$, and N_2 has finite uniform dimension. Let K_2 be a complement of $N_1 \oplus N_2$ in M. Note that Soc K_2 is essential in K_2, by the choice of N_1 and [1, Proposition 9.19].

If K_2 is not almost semisimple, then there exists a finitely generated semisimple submodule S_2 of K_2 such that S_2 is not a complement in K_2. Repeat this process. It produces a direct sum

$$N_1 \oplus N_2 \oplus N_3 \oplus \cdots$$

of submodules $N_i (i \geq 1)$ of M such that N_i has finite uniform dimension for all $i \geq 1$ and N_i is not semisimple for all $i \geq 2$. Because M is eventually semisimple, this process must stop, and hence K_t is almost semisimple for some $t \geq 1$.

Note that $L = N_1 \oplus \cdots \oplus N_t \oplus K_t$ is an essential submodule of M. Because K_t is a complement in M, it follows that L/K_t is an essential submodule of M/K_t. But $L/K_t \cong N_1 \oplus \cdots \oplus N_t$, which has finite uniform dimension. Thus M/K_t has finite uniform dimension.

Corollary 2.4. Let M be a locally Noetherian eventually semisimple module. Then there exists a semisimple complement K in M such that M/K has finite uniform dimension.

Proof. By Lemmas 2.2 and 2.3.

Lemma 2.5. Let M be a weak CS-module. Then every almost semisimple submodule of M is semisimple.

Proof. Let K be an almost semisimple submodule of M. Let $0 \neq x \in K$. Suppose that Soc xR is not finitely generated. Then

$$Soc\ xR = L_1 \oplus L_2 \oplus L_3 \oplus \cdots$$

for some inifinitely generated submodules $L_i (i \geq 1)$ of xR. For each $i \geq 1$, there
existes a direct summand N_i of M such that L_i is essential in N_i. Then the sum
$N_1 + N_2 + N_3 + ...$ is direct, and there exists a positive integer t such that N_t is
semisimple. In this case, $N_t = L_t$ so that L_t is a direct summand of M, and hence also
of xR. It follows that L_t is cyclic, a contradiction.

Thus Soc xR is finitely generated. By hypothesis, Soc xR is a direct summand
of xR. But Soc xR is essential in xR, and hence $xR = Soc\ xR$. It follows that K is
semisimple.

Lemmas 2.3 and 2.5 enable us to obtain the following extension of [2, Lemma 2].
Note that our inability to answer Question 1.4 means that the proof of the necessity
differs from that of [2].

Theroem 2.6. A module M is an eventually semisimple weak CS-module if and only
if $M = M_1 \oplus M_2$ for some semisimple module M_1 and weak CS-module M_2 with finite
uniform dimension.

Proof. Suppose that M is an eventually semisimple weak CS-module. By Lemmas
2.3 and 2.5, $M = M_1 \oplus M_2$ for some semisimple module M_1 and module M_2 with
finite uniform dimention. Let S be a semisimple submodule of M_2. Then $M_1 \oplus S$ is a
semisimple submodule of M. There exists a direct summand K of M such that $M_1 \oplus S$
is essential in K. Note that $K = M_1 \oplus (K \cap M_2)$, from which we infer that $K \cap M_2$
is a direct summand of M, and hence also of M_2. Furthermore, $S \subseteq K \cap M_2$. By [1,
Proposition 5.20], S is an essential submodule of the direct summand $K \cap M_2$ of M_2.
Thus M_2 is a weak CS-module.

Conversely, suppose that $M = M_1 \oplus M_2$ for some semisimple module M_1 and
weak CS-module M_2 with finite uniform dimension. By Lemma 2.1, M is eventually
semisimple, and , byu Lemma 1.10, a weak CS-module.

Corollary 2.7. Let M be a weak CS-module such that $M/(Soc\ M)$ has finite uniform
dimension. Then $M = M_1 \oplus M_2$ for some semisimple module M_1 and module M_2 with
finite uniform dimension.

Proof. By Theorem 2.6 and [2, Lemma 1].

Using Corollary 2.7, we can now generalize [2, Proposition 5].

Theorem 2.8. A weak CS-module M satisfies the ascending (respectively, descending) chain condition on essential submodules if and only if $M = M_1 \oplus M_2$ for some
semisimple module M_1 and Noetherian (respectively, Artinian) module M_2.

Proof. We give the proof in the Noetherian case; the proof in the Artinian case is
similar. If M is a direct sum of a semisimple module and a Noetherian module, then
M satisfies ACC on essential submodules, by [2, Lemma 4].

Conversely, suppose that M is a weak CS-module which satisfies ACC on essential submodules. By [2, Lemma 4], $M/(Soc\ M)$ is Noetherian. By Corollary 2.7, $M = M_1 \oplus M_2$ for some semisimple submodule M_1 and submodule M_2 with finite uniform dimension. There exist a positive integer k and uniform submodules $U_i (1 \le i \le k)$ of M_2 such that

$$E = U_1 \oplus \cdots \oplus U_k$$

is essential in M_2. Now M having ACC on essential submodules implies that U_i is Noetherian $(1 \le i \le k)$ and also that M_2/E is Noetherian. Thus M_2 is Noetherian.

3. THE OSOFSKY-SMITH THEOREM

Let R be a ring and M a right R-module. In this section we shall be concerned with the following theorem.

Theorem 3.1 (see [17]). Let M be a cyclic module such that N/K is a CS-module for every cyclic submodule N of M and submodule K of N. Then M has finite uniform dimension.

By adapting the proof, van Huynh, Dung and Wisbauer [9, Theorem 1.3] extended this theorem as follows:

Theorem 3.2. Let M be a cyclic module such that N/K is a direct sum of a CS-module and a module with finite uniform dimension for every cyclic submodule N of M and submodule K of N. Then M has finite uniform dimension.

In order to facilitate our discussion of generalizations of these theorems, we introduce certain operators on classes of modules. Let R be a ring. By a *class* X of R-modules we mean a collection of R-modules, containing a zero module, such that if $M \in X$ then $M' \in X$ for any module M' isomorphic to M. Particular classes of interest will be the classes

$$C,\ G,\ I,\ J,\ N\ \text{and}\quad U$$

of semisimple modules, finitely generated modules, injective modules, CS-modules, Noetherian modules and modules with finite uniform dimension, respectively.

Let X and Y be classes of modules. Then XY is the class of modules M which contain a submodule N such that $N \in X$ and $M/N \in Y$. Moreover, $X \oplus Y$ is the class of modules M such that $M = M_1 \oplus M_2$ for some submodules M_1 in X and M_2 in Y.

Let X be a class of modules. We define HX to be the class of modules M such that $M/N \in X$ for every submodule N of M. We define EX to be the class of modules M such that $M/N \in X$ for every essential submodule N of M. Finally, we define DX to be the class of modules M such that, for every submodule N of M, there exists a direct summand K of M such that $N \subseteq K$ and $K/N \in X$. For a discussion of these ideas, see [18] or [19]. For example, clearly $HN = N$, and it is proved in [18, Theorem 2.4 and Corollary 3.2], that

$$EN = CN \text{ and } \quad DN = C \oplus N .$$

It is proved in [19, Theorems 2.1 and 3.1], that

$$EU = C(HU) \text{ and } \quad DU = C \oplus (HU) .$$

Let X be a class of modules. We define CX to be the class of modules such that every cyclic submodule belongs to X. For example, CN is precisely the class of locally Noetherian modules. Given $A, B \in \{C, D, E, H\}$ we define

$$ABX = A(BX)) .$$

Lemma 3.3. Let X be any class of modules.

(i) $HX \subseteq DX \subseteq EX$

(ii) If $M \in HX$ (respectively, DX, EX), then so too does any homomorphic image of M.

(iii) If $M \in CX$ then so too does any submodule of M.

(iv) $CHX = HCX$.

(v) $CEX = ECX$.

Proof. (i), (iii) Clear.

(ii) See [18, Lemma 1.3].

(iv) The proof is straightforward. Note that $M \in CHX \iff mR \in HX$ for all $m \in M$
$\iff mR/N \in X$ for all $m \in M$ and each submodule N of mR.

On the other hand,

$M \in HCX \iff M/K \in CX$ for each submodule K of M
$\iff (xR + K)/K \in X$ for each submodule K and all $x \in M$
$\iff xR/(xR \cap K) \in X$ for each submodule K and all $x \in M$.

It easily follows that $CHX = HCX$.

(v) The proof of $CEX = ECX$ is similar. Note that if $m \in M$ and N is an essential submodule of mR then there exists an essential submodule K of M such that $K \cap mR = N$. For, let N' be a complement of N in M. Then $N \oplus N'$ is an essential submodule of M and $N' \cap mR = 0$. Thus $(N \oplus N') \cap mR = N$.

Question 3.4. Let X be a class of modules. Does $CDX = DCX$?

Using this notation, we can restate Theorem 3.2 thus:

Theorem 3.2'. Let M be a cyclic module such that $M \in CH(J \oplus U)$. Then $M \in HU$.

To see why the conclusion $M \in HU$ holds in Theorem 3.2′, we merely note that if N is any submodule of M, then, by Lemma 3.3, $M \in CH(J \oplus U)$ implies $M \in HC(J \oplus U)$, and hence $M/N \in HC(J \oplus U) = CH(J \oplus U)$. Thus, $M/N \in U$ by Theorem 3.2.

Now we shall extend Theorem 3.2′. First we shall prove a lemma.

Lemma 3.5. Let N be a submodule of a module M.

(i) If $N \in U$ and $MN \in U$, then $M \in U$.

(ii) If $N \in HU$ and $M/N \in HU$, then $M \in HU$.

Proof.

(i) Adapt the proof of Lemma 2.1.

(ii) Let K be a submodule of M. Then $M/(N + K) \in U$. On the other hand, $(N + K)/K \cong N/(N \cap K) \in U$. Thus, by (i), $M/K \in U$. It follows that $M \in HU$.

Corollary 3.6. Let M be a module and $M_i\,(1 \leq i \leq k)$ a finite collection of submodules of M such that $M = M_1 + \cdots + M_k$. Then $M \in HU$ if and only if $M_i \in HU$ for all $1 \leq i \leq k$.

Proof. Suppose first that $M \in HU$. Let $1 \leq i \leq k$. Let $N = M_i$. If K is a submodule of N, then N/K is a submodule of M/K. But $M/K \in U$. Thus $N/K \in U$. It follows that $N \in HU$.

Conversely, suppose that $M_i \in HU\,(1 \leq i \leq k)$. By induction on k, we can suppose without loss of generality that $k = 2$. But in this case $M/M_2 = (M_1 + M_2)/M_2 \cong M_1/(M_1 \cap M_2) \in HU$. Thus, $M \in HU$, by Lemma 3.5.

Combining Theorem 3.2′ and Corollary 3.6 we have the following generalization of Theorem 3.2′.

Theorem 3.7. For any ring R, $G \cap CH(J \oplus U) \subseteq HU$.

Our next aim is to generalize Theorem 3.7.

Lemma 3.8. For any ring R, $J \cap EU \subseteq C \oplus U$.

Proof. Let $M \in J \cap EU$. By [2, Lemma 9], $M/(Soc\ M) \in HU$. The result then follows by [2, Corollary 3].

An alternative proof of Lemma 3.8 can be given. Note that

$$J \cap EU \subseteq DU .$$

Then apply [19, theorem 3.1].

Corollary 3.9. For any ring R, $G \cap J \cap CE(J \oplus U) \subseteq U$.

Proof. Let $M \in G \cap J \cap CD(J \oplus U)$. Let N be any essential submodule of M. Let K be any submodule of M with $N \subseteq K$. Then K is an essential submodule of M. By Lemma 3.3, $CE(J \oplus U) = EC(J \oplus U)$, so that

$$M/K \in C(J \oplus U) .$$

Thus $M/N \in G \cap HC(J \oplus U)$. By Lemma 3.3 and Theorem 3.7, $N/M \in U$. Thus $M \in J \cap EU$. By Lemma 3.8, $M \in C \oplus U$. But M is finitely generated. We conclude $M \in U$.

We can now generalize Theorem 3.7.

Theorem 3.10. For any ring R, $G \cap CD(J \oplus U) \subseteq HU$.

Proof. Let $M \in G \cap CU(J \oplus U)$. Let $m \in M$. Then $mR \in D(J \oplus U)$. Let N be any submodule of mR. Then there exists a direct summand K of mR such that $N \subseteq K$ and $K/N \in J \oplus U$. Thus there exist submodules J and U of K, each containing N, such that

$$K/N = (J/N) \oplus (U/N) ,$$

$J/N \in J$ and $U/N \in U$.

By Lemma 3.3, $M \in CE(J \oplus U) = EC(J \oplus U)$, and hence $M/N \in EC(J \oplus U)$. Again applying Lemma 3.3, $M/N \in CE(J \oplus U)$, and hence $J/N \in CE(J \oplus U)$. Moreover, K, K/N and J/N are all cyclic. By Corollary 3.9, $J/N \in U$. Thus $K/N \in U$. It follows that $mR \in DU$. But $DU = C \oplus (HU)$, by [19, Theorem 3.1]. Hence $mR \in HU$, by Lemma 3.5.

We have proved that every cyclic submodule of M belongs to HU. Because M is finitely generated, $M \in HU$, by Lemma 3.5.

Corollary 3.11. Let R be a ring such that every cyclic right R-module is a direct sum of a projective module, a CS-module and a module with finite uniform dimension. Then every cyclic right R-module has finite uniform dimension.

Proof. Let $a \in R$. Let E be a right ideal of R such that $E \subseteq aR$. Then aR/E is a cyclic right R-module. By hypothesis, there exist right ideals $F_i (1 \leq i \leq 3)$ of R such that $E \subseteq F_i \subseteq aR(1 \leq i \leq 3)$, F_1/E is projective, F_2/E is a CS-module, F_3/E has finite uniform dimension, and

$$aR/E = (F_1/E) \oplus (F_2/E) \oplus (F_3/E) .$$

Note that $aR/(F_2 + F_3) \cong F_1/E$, so that $aR/(F_2 + F_3)$ is projective. Thus $(F_2 + F_3)$ is a direct summand of aR. It follows that $aR \in D(J \oplus U)$. Hence $R_R \in CD(J \oplus U)$. By Theorem 3.10, $R_R \in HU$. It follows that every cyclic right R-module has finite uniform dimension.

Let R be an arbitrary ring. For any ordinal $\alpha \geq 0$, K_α will denote the class of R-modules with Krull dimension at most α. In addition, L_α will denote the class of R-modules which contain an essential submodule with Krull dimension at most α. Note that $K_\alpha \subseteq L_\alpha$, for any ordinal $\alpha \geq 0$. For general facts about Krull dimension, see [5].

The next lemma is implicit in [5, pp. 16 and 33] (or see [8, Lemma 6]).

Lemma 3.12. For any ring R and ordinal $\alpha \geq 0$, $HL_\alpha = K_\alpha$.

Lemma 3.13. For any ring R and ordinal $\alpha \geq 0$,

$$G \cap CH(I \oplus L_\alpha) \subseteq K_\alpha \ .$$

Proof. Let $M \in G \cap CH(I \oplus L_\alpha)$. Then $M \in G \cap CD(J \oplus U)$, by Lemma 3.3 and [5, Proposition 1.4]. By Theorem 3.10, $M \in HU$. Let U be a cyclic uniform submodule of M. Let V be a maximal submodule of U. If $V = 0$ then U is simple. If $V \neq 0$, let $0 \neq v \in V$. Then $vR \in I \oplus L_\alpha$. But vR is indecomposable and not injective. Thus $vR \in L_\alpha$. It follows that $U \in L_\alpha$. Thus $M \in L_\alpha$.

Let N be any submodule of M. Then, by Lemma 3.3, $CH(I \oplus L_\alpha) = HC(I \oplus L_\alpha)$, and $M/N \in G \cap CH(I \oplus L_\alpha)$, so that, by the above argument, $M/N \in L_\alpha$. Thus $M \in HL_\alpha$. By Lemma 3.12, $M \in K_\alpha$.

Theorem 3.14. For any ring R and ordinal $\alpha \geq 0$,

$$G \cap CD(I \oplus L_\alpha) \subseteq K_{\alpha+1} \ .$$

Proof. Let $M \in G \cap CD(I \oplus L_\alpha)$. By Theorem 3.10 and [5, Proposition 1.4], $M \in HU$. Let N be any submodule of M. Then $M/N \in U$. Let U be a submodule of M containing N such that U/N is a cyclic uniform submodule of M/N. By Lemma 3.3,

$$M \in CE(I \oplus L_\alpha) = EC(I \oplus L_\alpha) \ ,$$

and hence $M/N \in EC(I \oplus L_\alpha) = CE(I \oplus L_\alpha)$, and finally $U/N \in CE(I \oplus L_\alpha) = EC(I \oplus L_\alpha)$.

Let V be any submodule of U properly containing N. Let W be any submodule of U properly containing V. Then W/N is an essential submodule of U/N, so that $U/W \in C(I \oplus L_\alpha)$. It follows that $U/V \in HC(I \oplus L_\alpha) = CH(I \oplus L_\alpha)$. By Lemma 3.13, $U/V \in K_\alpha$. Thus every proper homomorphic image of U/N belongs to K_α. Thus $U/N \in K_{\alpha+1}$. It follows that $M/N \in L_{\alpha+1}$. Thus $M \in HL_{\alpha+1}$. By Lemma 3.12, $M \in K_{\alpha+1}$.

Corollary 3.15. For any ring R and ordinal $\alpha \geq 0$,

$$CD(I \oplus L_\alpha) \subseteq CK_{\alpha+1} \, .$$

Proof. Let $M \in CD(I \oplus L_\alpha)$. Let N be any finitely generated submodule of M. Then $N \in CD(I \oplus L_\alpha)$ so that, by the theorem, $N \in K_{\alpha+1}$.

The next result generalizes [3, Theorem 4.1]. Compare it also with [8, Theorem 7].

Corollary 3.16. Let R be a ring and $\alpha \geq 0$ an ordinal. Suppose that every cyclic right R-module is a direct sum of a projective module, an injective module and a module containing an essential submodule of Krull dimension at most α. Then R has right Krull dimension at most $\alpha + 1$.

Proof. By Theorem 3.14, using the proof of Corollary 3.11.

It is proved in [17, Corollary 2], that for any ring R,

$$CHI \subseteq C \, .$$

Now we prove:

Theorem 3.17. For any ring R,

(i) $CEI \subseteq EC$, and

(ii) $CDI \subseteq CN$.

Proof.

(i) Let $M \in CEI$. Let N be an essential submodule of M. Let K be any submodule of M containing N. Then K is an essential submodule of M. By Lemma 3.3, $M \in ECI$, and hence $M/K \in CI$. It follows that $M/N \in HCI = CHI$. By [17, Corollary 2], $M/N \in C$. Thus $M \in EC$.

(ii) Let $M \in CDI$. Let N be any finitely generated submodule of M. Then $N \in CDI \subseteq CEI$, by Lemma 3.3. Thus $N \in EC$. Because N is finitely generated, it follows that $N \in EN$. Hence, by [18, Theorem 2.4], $N/(Soc\ N)$ is Noetherian. But $N \in G \cap CDI \subseteq G \cap CD(J \oplus U) \subseteq U$, by Theorem 3.10. Thus Soc N is Noetherian. It follows that N is Noetherian. Thus M is locally Noetherian.

REFERENCES

1. F.W. Anderson and K.R. Fuller, *Rings and categories of modules* (Springer-Verlag, 1974).
2. V. Camillo and M.F. Yousif, CS-modules with acc or dcc on essential submodules, preprint.

3. A.W. Chatters, A characterization of right Noetherian rings, Quart. J. Math. Oxford (2) **33** (1982) 65-69.

4. A.W. Chatters and C.R. Hajarnavis, Rings in which every complement right ideal is a direct summand, ibid., **28** (1977) 61-80.

5. R. Gordon and J.C. Robson, *Krull dimension*, Amer. Math. Soc. Memoirs 133 (1973).

6. M. Harada, On modules with extending property, Osaka J. Math. **19** (1982) 203-215.

7. M. Harada and K. Oshiro, On extending property of direct sums of uniform modules, Osaka J. Math. **18** (1981) 767-785.

8. D.V. Huynh, N.V. Dung and P.F. Smith, A characterization of rings with Krull dimension, J. Algebra, to appear.

9. D.V. Huynh, N.V. Dung and R. Wisbauer, A characterization of modules with finite uniform dimension, Arch. Math., to appear.

10. A.V. Jategaonkar, *Localication in Noetherian rings*, London Math. Soc. Lecture Note Series 98 (Cambridge University Press, 1986).

11. M.A. Kamal and B.J. Müller, Extending modules over commutative domains, Osaka J. Math. **25** (1988) 531-538.

12. M.A. Kamal and B.J. Müller, The structure of extending modules over Noetherian rings, ibid., 539-551.

13. M.A. Kamal and B.J. Müller, Torsion free extending modules, ibid., 825-832.

14. I. Kaplansky, Modules over Dedekind rings and valuation rings, Trans. Amer. Math. Soc. **72** (1952) 327-340.

15. S.H. Mohamed and B.J. Müller, *Continuous and discrete modules*, London Math. Soc. Lecture Note Series 147 (Cambridge University Press, 1990).

16. K. Oshiro, Continous modules and quasi-continuous modules, Osaka J. Math. **20** (1983) 337-372.

17. B.L. Osofsky and P.F. Smith, Cyclic modules whose quotients have complements direct summands, J. Algebra, to appear.

18. P.F. Smith, D.V. Huynh and N.V. Dung, A characterization of Noetherian modules, Quart. J. Math. Oxford, to appear.

19. P.F. Smith, Modules with many direct summands, Osaka J. Math., to appear.

20. R. Wisbauer, *Grundlagen der Modul- und Ringtheorie* (Verlag Reinhard Fischer, 1988).

Department of Mathematics
University of Glasgow
Glasgow, G12 8QW, Scotland UK

ON CONTINUOUS AND SINGULAR MODULES

S. Tariq Rizvi and Mohamed F. Yousif

1. INTRODUCTION

Rings for which every singular left module is injective, called SI-rings, were introduced and studied by Goodearl [4]. He characterized such rings as those nonsingular ones for which R/I is semisimple for every essential left ideal I of R. Continuous modules are an interesting generalization of (quasi-) injective modules (*e.g.*, see [5], [6], [8], [9]), and a study of a similar question in the context of continuous modules is of interest. In this paper, we investigate rings for which every singular left module is continuous. We call such a ring R, a left SC-ring (for singular is continuous) and show that R is characterized by the property that R/I is semisimple for every essential left ideal I of R. It is shown that this is equivalent to every (finitely-generated) singular left R-module to be (quasi-) continuous, This, in particular, settles an open question posed at the end of a recent paper of Osofsky and Smith [10]. We provide some characterizations of SC-rings, in particular, for semiprimary, for commutative and for regular rings. It is also shown that rings for which every finitely-generated (Goldie-) torsion module is (quasi-) continuous are precisely the SI-rings. Some results are also included for the case when every cyclic singular module is continuous (or CS).

2. DEFINITIONS AND PRELIMINARIES

All our rings have identity element and all modules are unital left modules (unless specified otherwise). The notations $J(R)$ and Soc $(_R R)$ (Soc (R_R)) stand for the Jacobson radical of the ring R and the left (right) socle of R, respectively. $Z_\ell(R)(Z_r(R))$ denote the left (right) singular ideal of R, $N \subset' M$ signifies that N is an essential submodule of M and a 'summand' means a direct summand. A module M is called singular if it is equal to its singular submodule $Z(M)$. M is N-injective if every homomorphism from a submodule of N into M can be extended to N.

Consider the following conditions on a module M :

(C_1) Every submodule of M is essential in a summand of M,

(C_2) Every submodule isomorphic to a summand of M is itself a summand,

(C_3) If M_1 and M_2 are summands of M with $M_1 \cap M_2 = 0$, then $M_1 \oplus M_2$ is a summand of M.

M is called continuous if it satisfies conditions (C_1) and (C_2), quasi-continuous if it satisfies (C_1) and (C_3), and a CS-module (also known as a "module with (C_1)" and "extending module") if M satisfies condition (C_1) only.

It is easy to see that $(C_2) \Rightarrow (C_3)$ and the hierarchy is as follows: injective $\Rightarrow$ quasi-injective $\Rightarrow$ continuous $\Rightarrow$ quasi-continuous $\Rightarrow$ CS; for more details we refer to [8].

A module M is called a V-module (GV-module) if every simple (simple singular R-module is M-injective. A ring R is a left V-ring (GV-ring) if $_R R$ is a V-module

(GV-module). R is known as a left S^3I-ring if every semisimple singular left R-module is injective (see [11]).

In the following, we list a few known results which will be used often.

Lemma 2.1 (Osofsky-Smith [10]) Let M be a cyclic module such that K/L is a CS-module for every cyclic submodule K of M and a submodule L of K. Then M is a finite direct sum of uniform modules.

Proof. See [10], Theorem 1.

Lemma 2.2 (Camillo-Yousif [2]) Let M be a cyclic module. If every cyclic singular subquotient of M is a CS-module, then every quotient of $M/Soc(M)$ is finite dimensional.

Proof. See [2], Lemma 11.

3. SC-RINGS

We first provide as a lemma the following basic result proved by Goodearl [4] (see also Yousif [14]).

Lemma 3.1. The following are equivalent for a ring R :

(1) R is a left SI-ring,

(2) $Z_\ell(R) = 0$ and every singular left R-module is semisimple,

(3) $Z_\ell(R) = 0$ and R/I is semisimple for every essential left ideal I of R.

Proof. See ([4], Proposition 3.1) and ([14], Proposition 2.5).

If we weaken the definition of a left SI-ring R, by requiring that every singular left R-module be quasi-injective, continuous or quasi-continuous, then it turns out (Theorem 3.2) that all of these conditions are, in fact, equivalent to the requirement every (cyclic) singular left R-module is semisimple. We call such a ring R, a left SC-ring.

Theorem 3.2. The following conditions are equivalent for a ring R :

(1) every singular left R-module is semisimple,

(2) every (finitely-generated) singular left R-module is quasi-injective,

(3) every (finitely-generated) singular left R-module is continuous,

(4) every (finitely-generated) singular left R-module is quasi-continuous,

(5) R/I is semisimple for every essential left ideal I of R.

Proof. It is easy to observe that the implications $(1) \Rightarrow (2) \Rightarrow (3) \Rightarrow (4)$ hold. We first prove $(4) \Rightarrow (5)$. Let every finitely-generated singular left R-module be quasi-continuous. If M is any cyclic singular left R-module, then every cyclic subquotient of M is singular, and hence by assumption a CS-module. By the Osofsky-Smith Theorem (Lemma 2.1), M is finite dimensional. Thus every cyclic singular left R-module is finite dimensional. Next, let L be any cyclic singular left R-module with $x \in L$, then $L \oplus Rx$ is quasi-continuous by assumption. Therefore, Rx is L-injective [8], and hence Rx is a summand of L. Thus L is finite dimensional and every cyclic submodule of L is a summand. Proposition 1.22 of [4], then yields that L is semisimple. Hence (5) holds.

$(5) \Rightarrow (1)$ is clear since if M is a singular left R-module, then $M = \sum_{x \in M} Rx$, and each Rx is semisimple by assumption. This completes the proof.

A ring R which satisfies the equivalent conditions of Theorem 3.2 is called a left SC-ring. A right SC-ring is similarly defined.

Next, we extend and generalize Propositions 3.3 and 3.6 of [4] for SC-rings in the following.

Corollary 3.3. If R is a left SC-ring, then

(i) $J(R) \subset Soc(_R R)$,

(ii) $J^2(R) = 0$,

(iii) $Z_\ell(R) \subseteq Soc(_R R)$,

(iv) $\bar{R} = R/Soc(_R R)$ is a left Noetherian, left V-ring, which is also left hereditary and left SI-ring.

Proof.

(i) Let R be a left SC-ring. Then by Theorem 3.2, R/I is semisimple for every essential left ideal I of R. Thus $J(R/I) = 0$, and so $J(R) \subseteq I$ for all $_R I \subset' {}_R R$. Hence $J(R) \subseteq Soc(_R R)$ holds.

(ii) $J^2(R) \subseteq J(R)Soc(_R R) = 0$.

(iii) As R/I is semisimple (hence Noetherian) for every essential left ideal I of R. Therefore, $\bar{R} = R/Soc(_R R)$ is Noetherian ([11], Corollary 2.26 and [3], Lemma 2). Also, for every $_R I \subset' {}_R R$, R/I is a left V-module (being semisimple), hence $\bar{R}$ is a left V-ring by ([15], Lemma 4). Thus $Z(_{\bar{R}} \bar{R}) = 0$ holds ([7]). Next, it is easy to check that a homomorphic image of a (left) SC-ring is a (left) SC-ring, hence $\bar{R}$ is a left SC-ring with $Z(_{\bar{R}} \bar{R}) = 0$. A combination of Theorem 3.2 and Lemma 3.1 yields that $\bar{R}$ is a left SI-ring. That $\bar{R}$ is left hereditary, follows directly from Proposition 3.3 of [4].

As noted in the preceding, nonsingular SC-rings are precisely the SI-rings, hence:

Proposition 3.4. For any ring R, the following are equivalent:

(1) $Z_\ell(R) = 0$ and R is a left SC-ring,

(2) R is a left SI-ring,

(3) Soc $(_R R)$ is projective and R is a left SC-ring.

Proof.

(1) $\Rightarrow$ (2) follows from Theorem 3.2 and Lemma 3.1.

(2) $\Rightarrow$ (1) is obvious, since left SI-rings are left nonsingular.

(3) $\Rightarrow$ (2). If R is a left SC-ring, then $R/Soc(_R R)$ is a left Noetherian, left V-ring by Corollary 3.3. Since Soc $(_R R)$ is also projective, it follows from ([11], Corollary 2.16) that R is a left $S^3 I$-ring and $Z_\ell(R) = 0$. Thus R is a left SI-ring.

(2) $\Rightarrow$ (3) is clear.

Remark. A semiprime left SC-ring is a left SI-ring. This holds, since every minimal left ideal is generated by an idempotent in a semiprime ring which implies that Soc $(_R R)$ is projective.

The following example shows that the SC-rings, in general, do not coincide with the SI-rings even in simple situations.

Example 3.5. Let $R = \mathbf{Z}/4\mathbf{Z}$, where $\mathbf{Z}$ is the ring of integers. It is easy to see that R has exactly three ideals $\{0, 2R, R\}$ and that R is an SC-ring (the only essential ideals are 2R and R itself). However, R is not an SI-ring, since the singular submodule $Z(R) = 2R$ is not injective.

Our next result provides a characterization of an SC-ring in terms of the SI-ring $R/Soc(R)$ in the following case.

Theorem 3.6. The following are equivalent for a ring R.

(1) R is a left SC-ring,

(2) $\bar{R} = R/Soc(_R R)$ is a left Noetherian, left SI-ring and left V-ring, such that $A/Soc(_R R)$ is semisimple for every left ideal A of R with $_R A\, ' \supset Soc(_R R)$.

Proof.

(1) $\Rightarrow$ (2) is clear by Corollary 3.3 and Theorem 3.2.

(2) $\Rightarrow$ (1). Set $S = Soc(_R R)$. To show that R is a left SC-ring, we claim that R/I is semisimple for every essential left ideal I of R : Suppose not, then let E be the maximal essential left ideal of R such that R/E is not semisimple. Let $- : R \longrightarrow R/S$ be the canonical map, as $\bar{R}/\bar{E} \cong R/E$ is not semisimple, we infer that $\bar{E}$ is not an essential left ideal of $\bar{R}$. Hence, there exists a nonzero left ideal $\bar{A}$ of $\bar{R}$, such that $\bar{E} \cap \bar{A} = 0$. This implies that $E \cap A = S$ where $S \subseteq A$ holds. It is easy to see that, in fact, $S \subset' A$: for, if there is a subideal B of A with $B \cap S = 0$, then $E \cap B \subseteq E \cap A = S$ and $E \cap B \subset S \cap B = 0$ holds. But $E \subset' R$, thus $B = 0$ and hence $S \subset' A$. Therefore, by hypothesis A/S is semisimple as a left R-module.

Since R/E is uniform, and as $A/S = A/A \cap E \cong (A + E)/E \subset R/E$, therefore, A/S is simple as a left $\bar{R}$-module. Since $\bar{R}$ is a left V-ring, this yields that A/S is R-injective and R/E is simple as a left $\bar{R}$-module, and hence as a left R-module. This is a contradiction. Therefore, R is a left SC-ring.

Corollary 3.7. The following conditions are equivalent for a ring R.

(1) R is a left SC-ring with $R' \supset Soc(_R R)$,

(2) $\bar{R} = R/Soc(_R R)$ is semisimple artinian.

Proof.

(1) $\Rightarrow$ (2) is obvious by Theorem 3.6 (or Theorem 3.2).

(2) $\Rightarrow$ (1). If $\bar{R}$ is semisimple artinian, then the statement (2) in Theorem 3.6 is automatically satisfied, and hence R is a left SC-ring. To show that $R' \supset Soc(_R R)$, set $S = Soc(_R R)$ and let $_R I$ be a nonzero left ideal with $I \cap S = 0$. Then $I = I/I \cap S \cong I + S/S \subset R/S$ holds and hence $_R I$ is semisimple. Therefore, $_R I \subset S$, a contradiction.

If R is a commutative ring, then by (Theorem 3.9, [4]), R is an SI-ring if and only if R is regular and $R/Soc(R)$ is semisimple. Our next result provides a similar characterization of SC-rings for the commutative case.

Theorem 3.8 (cf [1], Theorem 2). Let R be a commutative ring. Then R is an SC-ring if and only if $\bar{R} = R/Soc(R)$ is semisimple artinian.

Proof. Let R be an SC-ring. Then $\bar{R}$ is a Noetherian V-ring. As R is commutative, $\bar{R}$ is regular and hence semisimple artinian. The converse follows from Corollary 3.7.

Proposition 3.9. Let R be a left SC-ring.

(i) If $_R R$ is a CS-module, then R is left Noetherian.

(ii) If $_R R$ is continuous, then R is left artinian.

Proof.

(i) As R is a left SC-ring, R/I is semisimple for every essential left ideal I of R. Therefore, by ([12], Proposition 4.6) every cyclic left R-module is a direct sum of a projective module and a semisimple module. Then ([12], Theorem 1.5) yields that R is left Noetherian.

(ii) If R is a left continuous, left SC-ring, then it is left Noetherian by the preceding. The result from [5] then implies that R is left artinian.

A module M is called torsion (or Goldie torsion) if $M = Z_2(M)$ where $Z_2(M)/Z(M)$ $Z(M/Z(M))$. It is well known that, in general, every singular module is torsion, and if the ring is non-singular, then the class of torsion modules coincides with that of the singular modules. Our next result generalizes ([1], Theorem 2).

Theorem 3.10. The following are equivalent for a ring R:

(1) every (finitely-generated) torsion left R-module is continuous,

(2) every (finitely-generated) torsion left R-module is quasi-continuous,

(3) R is a left SI-ring.

Proof.

(1) $\Rightarrow$ (2) is obvious.

We first prove that (2) $\Rightarrow$ (3). Assume that every finitely-generated torsion left R-module is quasi-continuous. Let Ry be any cyclic torsion R-module, and let $x \in Ry$. Then $Rx \oplus Ry$ is quasi-continuous and, therefore, Rx is Ry-injective. This implies that Rx is a summand of Ry and hence every cyclic submodule of Ry is a summand of Ry. As every cyclic submodule of Ry is quasi-continuous (hence a CS-module), Ry is finite dimensional by Lemma 2.2. Therefore, Ry is semisimple. Thus every torsion left R-module is semisimple. Next, let S be any singular left R-module. Then $E(S)$ is torsion and hence semisimple, so $E(S) = S$ holds. Therefore, R is a left SI-ring.

(3) $\Rightarrow$ (1). Let R be a left SI-ring. Then R is left nonsingular and, therefore, $Z_2(M) = Z(M)$ holds. An application of Theorem 3.2 now yields the result.

Osofsky and Smith have shown that if every cyclic singular module is injective then the ring is an SI-ring ([10], Corollary 5). However, if we only require that every cyclic singular module be continuous, then the ring may not be even an SC-ring. Specifically, the following example shows that if every cyclic torsion (singular) module is only continuous, then every finitely-generated torsion (singular) module need not be continuous, as the ring need not be an SI-ring, and the preceding theorem fails to hold.

Example 3.11. Let $R = \mathbf{Z}$, the ring of integers. Then $\mathbf{Z}$ is a nonsingular ring and every cyclic torsion (= singular) $\mathbf{Z}$-module is quasi-injective. However, it is a non-regular, non-splitting ring which is not an SI-ring (and as $\mathbf{Z}$ is nonsingular, also not an SC-ring). Clearly, not every finitely-generated torsion $\mathbf{Z}$-module is quasi-injective (or semisimple or quasi-continuous). Since otherwise every singular $\mathbf{Z}$-module will be semisimple, while $\mathbf{Z}_n$ is a singular non-semisimple $\mathbf{Z}$-module.

In the remainder of this section, we consider the rings for which every cyclic singular module is a continuous or a CS-module, and provide characterizations in some special cases.

Recall that for a regular ring R, the SI-property on the left is equivalent to the SI-property on the right, and it is characterized by $R/Soc(R)$ to be semisimple artinian ([4], Corollary 3.7). Our next observation is an analog of this result for SC-rings.

Proposition 3.12. For a regular ring R, the following are equivalent:

(1) R is a left SC-ring,

(2) every cyclic singular left (or right) R-module is a CS-module,

(3) R is a right SC-ring.

Proof. We note that for a regular ring R, $Z(R) = 0$ and Soc $(_R R) = Soc(R_R)$ holds. Therefore, in view of Proposition 3.4, a left SC-ring is equivalent to a left SI-ring. Thus the equivalence between (1) and (3) follows directly from ([4], Corollary 3.7). We show (2) $\Rightarrow$ (1) : If every cyclic singular left (or right) R-module is a CS-module, then by Lemma 2.2, $R/Soc(R)$ is finite dimensional. As R is regular, $R/Soc(R)$ is semisimple artinian, hence by ([4], Corollary 3.7) R is a left SI- (and, therefore, a left SC-) ring.

(1) $\Rightarrow$ (2) is obvious.

Example 3.13. The Example 3.8 in [4], page 52, also exhibits that, in general, a right SC-ring need not be a left SC-ring (use Proposition 3.4).

Proposition 3.14. Let R be a left GV-ring. If every cyclic singular left R-module is a CS-module, then R is a left $S^3 I$-ring.

Proof. Let every cyclic singular left R-module be CS. Then Lemma 2.2 yields that every quotient of $\bar{R} = R/Soc(_R R)$ is finite dimensional. As R is a left GV-ring, $\bar{R}$ is a left V-ring and hence every left $\bar{R}$-module is a V-module. Therefore, every nonzero left R-module as a maximal submodule. Now as every quotient of $\bar{R}$ is finite dimensional, Theorem 3.8 of [13] then yields that $_{\bar{R}}\bar{R}$ is Noetherian. Therefore, since R is a left GV-ring and since $R/Soc(_R R)$ is left Noetherian, by ([11], Corollary 2.16), it follows that R is a left $S^3 I$-ring.

Jain and Mohamed [6] call a ring R a left CC-ring if every cyclic left R-module is continuous. The next observation for continuous modules is similar to Proposition 2 of [1] for quasi-injective ones.

Proposition 3.15. Let $Soc(_R R) \subset' R$. Then every cyclic singular left R-module is continuous if and only if $\bar{R} = R/Soc(_R R)$ is a left CC-ring.

Proof. Let every cyclic singular left R-module be continuous and let $\bar{I} \subset \bar{R}$ be any left $\bar{R}$-ideal. As Soc $(_R R) \subset I$ and Soc $(_R R) \subset' {_R R}$, thus $_R I \subset' {_R R}$. Therefore, R/I is continuous. But $R/I \cong \bar{R}/\bar{I}$, and hence $\bar{R}/\bar{I}$ is continuous as a left $\bar{R}$-module. Therefore, $\bar{R}$ is a left CC-ring.

Conversely, let $\bar{R}$ be a left CC-ring and let $M = R/E$ be any cyclic singular left R-module, where $_R E \subset' {_R R}$. Then $M = R/E$ is a cyclic left $\bar{R}$-module and hence continuous as an $\bar{R}$ (therefore, as an $R-$) module.

Our final theorem is a result for continuous modules analogous to ([1], Theorem 8) proved by Ahsan for quasi-injectives. The proof of our theorem also provides a simple short proof of Ahsan's result.

Recall that a ring R is left duo if every left ideal is a two-sided ideal.

Theorem 3.16. Let R be a left duo ring, then the following are equivalent.

(1) R is a left continuous ring such that every cyclic singular left R-module is continuous,

(2) R is a left CC-ring.

Proof. Let R be left continuous and let every cyclic singular left R-module be continuous. If I is any left ideal of R, then $I \subset' Re$, where $e^2 = e$ and $R = Re \oplus R(1 - e)$. Thus $R/I \cong Re/I \times R(1 - e)$ holds. Since R is left duo, this is a ring decomposition, hence Re/I and $R(1 - e)$ are relatively injective to each other (there are no nontrivial maps between them). Next, as Re/I is singular, it is continuous by hypothesis, and $R(1 - e)$ is continuous being a summand of a left continuous ring. An application of ([9], Theorem 13) then implies that R/I is left continuous. Thus R is left CC-ring.

The converse is obvious.

Remark. Our last few results provide some information; however, it is still open to characterize arbitrary rings for which every cyclic singular left module is a continuous (or a CS-) module. Also open is to characterize rings for which every singular is a CS-module.

Acknowledgements. The authors thank Professor B.J. Müller for many helpful discussions and for his help with Theorem 3.6. Partial support received from The Ohio State University, Lima, in the form of a Faculty Professional Leave and a Research grant to the first author and a Special Research Assignment quarter to the second author, is gratefully acknowledged.

REFERENCES

1. J. Ahsan, Rings characterized by their torsion cyclic modules, Houston J. Math (**14**), No. 3 (1988) 291-303.
2. V. Camillo and M.F. Yousif, CS-modules with acc or dcc on essential submodules (to appear in Comm. in Algeb.).
3. N.V. Dung, D.V. Huynh and R. Wisbauer, Quasi-injective modules with acc or dcc on essential submodules, Arch. Math (to appear).
4. K.R. Goodearl, Singular torsion and the splitting properties, Memoirs of the Amer. Math. Soc., No. **124** (1972).
5. S.K. Jain, S.R. López-Permouth and S.T. Rizvi, Continuous rings with acc on essentials are artinian, Proc. Amer. Math. Soc., Vol. **108**, No. 3 (1990) 583-586.
6. S.K. Jain and S. Mohamed, Rings whose cyclic modules are continuous, J. Indian Math. Soc. **42** (1978) 197-202.
7. G.O. Michler and O.E. Villamayor, On rings whose simple modules are injective, J. Algeb. **25** (1973) 185-201.
8. S.H. Mohamed and B.J. Müller, *Continuous and discrete modules*, London Math. Soc. Lecture Note Series **147**, Cambridge University Press (1990).
9. B.J. Müller and S.T. Rizvi, On injective and quasi-continuous modules, J. Pure and App. Algeb. **28** (1983) 197-262.
10. B.L. Osofsky and P.F. Smith, Cyclic modules whose quotients have complements direct summands (to appear in J. Algeb.).

11. S. Page and M.F. Yousif, Relative injectivity and chain conditions, Comm. Algeb. **17** (4) (1989) 899-924.

12. P.F. Smith and D.V. Huynh, Characterising rings by their modules, Banach Centre Publications (Warsaw) **26** (1989).

13. R.C. Shock, Dual generalizations of the artinian and Noetherian conditions, Pacific J. Math., Vol. **54**, No. 2 (1974) 227-235.

14. M.F. Yousif, SI-modules, Math. J. Okayama Univ. **28** (1986) 133-146.

15. _____ , V-modules with Krull dimension, Bull. Austral. Math. Soc., Vol. **37** (1988) 237-240.

Department of Mathematics
The Ohio State University
Lima, Ohio 45804

PERMUTATION IDENTITY RINGS AND THE MEDIAL RADICAL

Gary Birkenmeier and Henry Heatherly

1. INTRODUCTION

A ring R is a *permutation identity ring* if for some $n \geq 2$, there exists a permutation σ, $\sigma \neq 1$, on n symbols, such that $a_1 a_2 \cdots a_n = a_{\sigma 1} a_{\sigma 2} \cdots a_{\sigma n}$, for each a_1, $\cdots$, $a_n \in R$. The class *Perm* of permutation identity rings is a proper subclass of the class of polynomial identity (P.I.) rings. The class *Perm* contains much more than the obvious commutative rings, nilpotent rings and their finite direct sums. We will illustrate the diversity in the makeup of *Perm* with examples and construction methods. Medial rings, those satisfying the identity: abcd = acbd, play a key role in our study of permutation identity rings. Also of interest are *left permutable rings*, those satisfying abc = bac identically. Medial or left permutable rings have been the subject of several recent works [3,4,16]. The general permutation identity setting for rings seems to have only been considered heretofore by Nordahl [15], where he discussed semigroup algebras which are permutation identity rings. (Nordahl calls these objects "permutative semigroup algebras".) Analogous to the definitions above for rings, one can define *permutation identity semigroups, medial semigroups*, etc. and similarly for many other algebraic systems. Both historically and from the vantage point of our personal introduction to the study, medial semigroups serve as the principal motivation for studying other medial algebraic systems. There is an extensive literature on mediality, left (or right) permutability, and other permutation identities for semigroups and groupoids. (For an extensive list of references, see [5,10,11,14].) Collaterally, there has been recent rising interest in near-rings which satisfy various permutation identities. (See [2] for further references and details.)

In this paper we develop the basic properties of permutation identity rings, observe their behavior under various chain conditions, and develop a radical property connected with the class of permutation identity rings. If R is a permutation identity ring, then the ideal generated by the set $[R, R] = \{ab - ba : a, b \in R\}$, denoted by $< R, R >$, is nilpotent; the set $N(R)$ of all nilpotent elements of R is an ideal of R. Thus the Levitzki radical $\mathcal{L}(R)$, the upper nil radical $\mathcal{N}(R)$, and $N(R)$ are all equal. The Jacobson radical $J(R)$ is characterized as the intersection of all ideals M such that R/M is a field. We show that every prime (semiprime) ideal of R is completely prime (semiprime). Consequently, the prime radical $\mathcal{P}(R)$ is equal to $N(R)$. If R satisfies either A.C.C. on nilpotent ideals or D.C.C. on ideals, then every nil subring of R is nilpotent. This is a much sharper result than obtainable for P.I. rings in general. A quasiradical is given which determines a radical property $\bar{M}$, it is shown that the upper radical property determined by the class of commutative rings is the same as that determined by the class of permutation identity rings, and this upper radical is $\bar{M}$. Further properties of this radical and some associated ideals are given, as well as some examples which serve to illustrate and delimit the theory.

Some remarks on notation are in order. For an arbitrary ring A, if S is a nonempty subset of A, then $< S >_A$ will denote the ideal in A generated by S. We use $l_A(S)$

and $r_A(S)$ for the left and right annihilators of S, respectively. The full ring of n by n matrices over A is denoted by $A_{n \times n}$. If no ambiguity will arise, we use the symbols, N, $\mathcal{N}$, $< S >$, $l(S)$, etc. without the identifying symbol for the ring with respect to which these sets are formed.

2. BASIC RESULTS AND CHAIN CONDITIONS

Denote by *Perm* σ the class of all rings which satisfy a fixed permutation identity σ. Note that *Perm* and *Perm* σ each are closed under the formation of subrings, factor rings, finite direct sums, and polynomial rings and formal power series rings in any number of commuting indeterminants. Each *Perm* σ is closed under direct product. We shall see in the sequel that the classes *Perm* σ and *Perm* are not closed under the formation of full matrix rings. A nilpotent ring of index n is a permutation identity ring which satisfies a permutation identity on n or less elements. More generally, for any ring A, if A^n is contained in the center of A, then A satisfies every permutation identity on k elements, for $k \geq n + 1$.

If R is a free ring and I is the ideal of R generated by the set $\{(a_1 a_2 \cdots a_n) - (a_{\sigma 1} a_{\sigma 2} \cdots a_{\sigma n}) : a_1, a_2, \cdots, a_n \in R\}$, then R/I is a permutation identity ring satisfying the identity induced by the permutation σ.

To better indicate the diversity of the class of permutation identity rings, we give a selected list of examples and methods of constructing examples of such rings in addition to those mentioned above.

We say a permutation identity P_0 is *stronger* than the permutation identity P_1 if whenever a ring satisfies P_0 it also satisfies P_1. (Similarly, P_1 is *weaker* than P_0.) For example, left permutability is stronger than mediality.

Example 2.1. Let S be a multiplicative semigroup which satisfies a permutation identity and let R be a commutative ring (not necessarily with unity). Then the semigroup ring $R[S]$ satisfies the same permutation identity that S does. More generally, if R is a ring satisfying a permutation identity P_0 and S satisfies a permutation identity P_1, with P_0 stronger than P_1, then $R[S]$ satisfies P_1.

Since a group which satisfies a permutation identity is commutative, the only group rings arising in the above construction are those of the form $R[S]$, where S is commutative. In this case $R[S]$ satisfies the same permutation identity that R does.

Example 2.2. Let R be a ring which satisfies a permutation identity given by the permutation σ on n letters. Let M be a left R-module with R-homomorphisms $f : M \to R$, $h : M \to M$ satisfying $h^2 = h$ and $fh = f$. Define $a * b = f(a)h(b)$, for each a, $b \in M$. Then $(M, +, *)$ is a ring and it satisfies a permutation identity on $n + 1$ letters. (For example, if R is right permutable, then $(M, +, *)$ is medial.)

This construction is a special case of a much more general one given in [1]. Many specific examples yielding medial and left permutable rings from the construction in Example 2.2 are discussed in [3].

A somewhat similar process to that of Example 2.2 is to use any ring T which has the following properties: (1) there exists a nontrivial permutation σ, on n letters; (2) there exists a nonzero element $c \in R$ such that $c \in l(A_\sigma)$, where $A_\sigma = \{(a_1 a_2 \cdots a_n) - (a_{\sigma 1} a_{\sigma 2} \cdots a_{\sigma n}) : a_1, \cdots, a_n \in T\}$. Then define $a * b = acb$. The resulting ring $(T, +, *)$ satisfies a permutation identity on $n + 2$ letters.

Example 2.3. Let M be a free module over a commutative ring R with free basis $\{b_1, \cdots, b_n\} = B$. Define $b_i \cdot b_j = \sum \lambda_{ijk} b_k$, $k = 1, \cdots, n$. This extends linearly to all of M and yields a (not necessarily associative) algebra over R. If the structure constants λ_{ijk} are chosen so that $(b_i \cdot b_j) \cdot b_m = b_i \cdot (b_j \cdot b_m)$, then the algebra W will be associative. If also the λ_{ijk} are chosen so that the set B satisfies a permutation identity, then $(W, \cdot)$ will satisfy that permutation identity as well.

If S is a semigroup which satisfies a permutation identity, then Putcha and Yaqub [17] have shown that there is an integer $n \geq 1$ so that $uxyv = uyxv$, for each $u, v \in S^n$, $x, y \in S$. From this it follows that if R is a ring which satisfies a permutation identity, then we have the identity:

$$uxyv = uyxv \text{ , for each } x, y \in R , u, v \in R^n . \qquad (1)$$

Conversely, if R is a ring satisfying this identity for some n, then R satisfies a permutation identity. The smallest m for which identity (1) holds will be called the *Putcha-Yaqub number* for R. For completeness we say a commutative ring has Putcha-Yaqub number zero. It is worth noting that a ring can satisfy a permutation identity on three element sets yet have a Putcha-Yaqub number of one (*e.g.*, left permutable rings). If $\{R_\lambda : \lambda \in \Lambda\}$ is a set of permutation identity rings and there is a fixed n such that the Putcha-Yaqub number of each R_λ is no larger than n, then ΠR_λ, $\lambda \in \Lambda$, has Putcha-Yaqub number $\leq n$.

Let R be a ring which satisfies a permutation identity, say with Putcha-Yaqub number n, then $R^n < R, R > R^n = 0$ and hence $< R, R >$ is nilpotent of index $\leq 2n + 1$. It is well known that in any ring in which there is a fixed n so that $(ab - ba)^n = 0$, for all a, b, the set of nilpotent elements is an ideal and this set is equal to be Levitzki radical [6, p. 131]. So $\mathcal{L}(R) = \mathcal{N}(R) = N(R)$. Any nil subring of R is locally nilpotent. This follows from [8, Corollary 1, p. 91]; however, it is easy to obtain directly in the context of a permutation identity ring by using that for any nil subring T of R which is finitely generated, the factor ring $T/ < T, T >$ is commutative and locally nilpotent.

In the class of permutation identity rings the Köthe Conjecture is resolved affirmatively. This suggests the following more general result.

Proposition 2.4. If A is a ring in which $< A, A >$ is nil, then every nil one-sided ideal of A is contained in a nil ideal.

Proof. If Y is a nil left ideal in A, then Y maps onto a nil ideal $\phi(Y)$ in the commutative ring $A/ < A, A >$, under the natural epimorphism $\phi : A \to A/ < A, A >$. Let x be an arbitrary element in the ideal $\phi^{-1}(\phi(Y))$. Then $x+ < A, A >\in \phi(Y)$. So

$(x+ <A, A >)^n = 0$, for some n, and hence $x^n \in< A, A >$, yielding x is nilpotent and hence $\phi^{-1}(\phi(Y))$ is a nil ideal containing Y. Similarly for right ideals of $\mathbf{A}$.

Proposition 2.4 extends a result we obtained previously for medial rings [3], where a different proof was given, and also yields some results due to Nordahl as a corollary.

Corollary 2.5 (Nordahl [15]). If S is a semigroup which satisfies a permutation identity and F is an arbitrary field, then for the semigroup algebra $F[S]$:

(1) $N(F[S]) = \mathcal{N} (F[S])$;

(2) if $J(F[S]) = 0$, then $F[S]$ is commutative.

For many important classes of rings, those which satisfy a permutation identity must be commutative. In light of $R^n < R$, $R > R^n = 0$, we see that if R satisfies a permutation identity and any one of the following, then R is commutative:

(1) R semiprime (and hence semi-simple with respect to most of the standard radicals),

(2) R contains an element which is not a left zero divisor and an element which is not a right zero divisor,

(3) R has unity element.

In particular, if R is simple, then either $R^2 = 0$ or R is a field. While if R has a left (right) unity element, then R is left (right) permutable.

We use these observations in establishing results about maximal, prime and semiprime ideals in permutation identity rings.

Proposition 2.6. If R is a permutation identity ring and M is a maximal ideal of R, then either (exclusively):

(1) R/M is a field, M is a completely prime ideal of R, and M is maximal as a left or right ideal of R; or

(2) $R^2 \subseteq M$, $(R/M)^+ \approx C_p$, p a prime and M is maximal as a subgroup of R^+.

Note that if R has a left or right unity, then (1) holds.

Proposition 2.7. Let R be a permutation identity ring.

(1) If R is a right (left) primitive ring, then R is a field.

(2) If I is a right (left) primitive two-sided ideal, then R/I is a field (and hence I is a maximal and a prime ideal).

(3) $J(R)$ is equal to the intersection of all ideals M of R such that R/M is a field. (If there are no such ideals, then $J(R) = R$.)

(4) If $J(R) = 0$, then R is isomorphic to a subdirect product of fields.

Proof.

(1) Right primitive rings are prime rings; so R is commutative and hence is a field.

(2) This is immediate from (1).

(3) and (4) $J(R)$ is the intersection of all right primitive two-sided ideals of R [6, Theorem 34]; so if $J(R) = 0$, then R is a subdirect product of fields.

Note that in a permutation identity ring, left and right primitive are equivalent.

Proposition 2.8. Let P be an ideal in a permutation identity ring R. The following are equivalent:

(1) P is a prime ideal;

(2) R/P is a commutative domain;

(3) P is a completely prime ideal.

Proof. (1) $\Rightarrow$ (2). The prime ring R/P is commutative; so R/P is a domain.

The other implications are immediate.

Proposition 2.9. Let S be an ideal in a permutation identity ring R. The following are equivalent:

(1) S is a semiprime ideal;

(2) R/S is reduced;

(3) S is a completely semiprime ideal.

Proof. If S is a semiprime ideal, then R/S is a semiprime permutation identity ring and hence is commutative. So R/S is reduced. The rest is immediate.

Corollary 2.10. If R is a permutation identity ring, then $P(R) = N(R)$.

Proof. In general $P(R) \subseteq N(R)$. Since $P(R)$ is a semiprime ideal, then $R/P(R)$ is reduced and hence $N(R) \subseteq P(R)$.

Proposition 2.11. Let R be a ring which satisfies a permutation identity.

(1) If I is a minimal ideal of R, then either $I^2 = 0$ or I is a field.

(2) If R is subdirectly irreducible with heart H, then either $H^2 = 0$ or $R = H$ is a field.

Proof. A minimal ideal of a ring is either square zero or is a simple ring [6, p. 75]. So $I^2 = 0$ or I is a simple medial ring, which is not square zero, *i.e.*, a field. Thus if R is subdirectly irreducible, H is either square zero or is a field. It is well known that if the heart of a ring is itself a ring with unity, then the heart is equal to the whole ring.

The question of when nil (left, right, two-sided) ideals are nilpotent is one which has been approached from many avenues, including that of P.I. rings or algebras. Various types of finiteness conditions have been invoked. The special properties of permutation identity rings enable us to obtain "nil implies nilpotent" under weaker chain conditions than those for arbitrary rings.

Proposition 2.12. Let R be a permutation identity ring. If R satisfies the A.C.C. on nilpotent ideals, then every nil subring of R is nilpotent.

Proof. We use the well-known fact that a commutative ring with A.C.C. on nilpotent ideals has every nil subring nilpotent. Any nil subring T of R maps onto the nil subring $\bar{T}$ of $R/<R,R>$. Since $<R,R>$ is nilpotent, the A.C.C. on nilpotent ideals is inherited in the commutative ring $R/<R,R>$. So $\bar{T}$ is nilpotent, say $(\bar{T})^n = 0$, and hence $T^n \subseteq <R,R>$. Thus T is nilpotent.

The proof given here is different from the one we gave for medial rings [3]. In fact, the above proof scheme shows the following: if I is a nilpotent ideal of a ring A and A/I has the property that each nil subring is nilpotent, then each nil subring of A is nilpotent. An interesting special case of this is where $I = <A, A>$ and A/I has A.C.C. on nilpotent ideals.

Proposition 2.13. Let R be a ring which satisfies a permutation identity. If R has D.C.C. on ideals, then:

(1) Every nil subring of R is nilpotent;

(2) Every prime ideal P of R is maximal and R/P is a field;

(3) $J(R) = N(R) = \mathcal{P}(R)$;

(4) If $R^2 = R$, then the set of maximal and of prime ideals coincide;

(5) The number of prime ideals of R is finite (so the structure space of R is finite).

Proof. The commutative ring $R/<R,R>$ has D.C.C. on ideals, and hence every nil ideal of $R/<R,R>$ is nilpotent. Thus the image X of $\mathcal{N}(R)$ in $R/<R,R>$ is nilpotent, say $X^m = 0$. Consequently $(\mathcal{N}(R))^m \subseteq <R,R>$ and hence $\mathcal{N}(R)$ is nilpotent. Since $N(R) = \mathcal{N}(R)$, every nil subring of R is contained in $\mathcal{N}(R)$, so each such subring is nilpotent.

Since R/P is a commutative domain (Proposition 2.8) and has D.C.C. on ideals, then R/P is a field and P is maximal. From $J(R)/N \subseteq J(R/N) = 0$, we have $J(R) = N$.

If $R^2 = R$, then by Proposition 2.6 each maximal ideal is prime. Finally, because for a prime ideal P we have R/P is a field, the standard commutative ring proof can be used, *mutatis mutandis*, to obtain that there are only finitely many prime ideals of R.

(Note: For part (1) of Proposition 2.13, it is sufficient to assume only D.C.C. on (two-sided) ideals of the form xN, with $x \in N$.)

Propositions 2.12 and 2.13 extend to the class of permutation identity rings results given previously for commutative rings. The chain conditions used here are different than those used in a more general setting by Levitzki [13], Herstein and Small [9] or Fisher [7]. Propositions 2.12 and 2.13 gain increased interest in light of the following example, which is of a medial ring satisfying both A.C.C. and D.C.C. on ideals, but which satisfies neither chain condition on either left or right ideals.

Example 2.14. Let K be the real algebra defined via the basis $\{x,y\}$ for $\mathbf{R}^2$ by $x^2 = x$, $yx = y$, $xy = y^2 = 0$. This yields a ring in which every proper, nonzero right ideal is a one-dimensional subspace and which has neither D.C.C. nor A.C.C. on left ideals (see Divinsky [6, Ex. 5, p. 37]). This ring is right permutable. Then the ring $R = K \oplus K^{opp}$ is a medial ring with D.C.C. and A.C.C. on ideals, but with neither chain condition on left or on right ideals.

The following shows that the only matrix rings which are permutation identity rings are those which are nilpotent.

Proposition 2.15. Let R be the ring of all n by n matrices, $n \geq 2$, over a ring A. If R satisfies a permutation identity, then $A^{2m+2} = 0$, where m is the Putcha-Yaqub number of R.

Proof. The idea of the proof is conveyed fully in the case where $n = 2$. Let a_j, d_j, b, $c \in A$, $j = 1, 2, \ldots, m$. Then

$$q = \Pi \begin{bmatrix} 0 & 0 \\ 0 & a_j \end{bmatrix} = \begin{bmatrix} 0 & 0 \\ 0 & \Pi a_j \end{bmatrix}, \ p = \Pi \begin{bmatrix} d_j & 0 \\ 0 & 0 \end{bmatrix} = \begin{bmatrix} \Pi d_j & 0 \\ 0 & 0 \end{bmatrix},$$

where these products range over $1, \ldots, m$, and hence

$$q \cdot \begin{bmatrix} 0 & 0 \\ b & 0 \end{bmatrix} \begin{bmatrix} c & 0 \\ 0 & 0 \end{bmatrix} \cdot p = \begin{bmatrix} 0 & 0 \\ (\Pi a_j)(bc)\Pi d_j & 0 \end{bmatrix}.$$

Also, $q \cdot \begin{bmatrix} c & 0 \\ 0 & 0 \end{bmatrix} \begin{bmatrix} 0 & 0 \\ b & 0 \end{bmatrix} \cdot p = \begin{bmatrix} 0 & 0 \\ 0 & 0 \end{bmatrix}$. So $(\Pi a_j)(bc)(\Pi d_j) = 0$.

Note that the nilpotent index of A may be less than $2m + 2$. For example, if A is left permutable, then a similar argument to the above gives $A^3 = 0$.

Similar results to that of Proposition 2.15 hold for upper triangular and lower triangular matrix rings.

There are many P.I. rings without unity which are not permutation identity rings. Some P.I. rings are given by a polynomial identity which is not a permutation identity, yet the ring does satisfy a permutation identity. This can be trivial to see, as in the case of nilpotent rings, or subtle as in the case of certain commutativity theorems and in the following result.

Proposition 2.16. If R is a ring which satisfies the identity: $abc = abac$, then R is left permutable.

For a proof and for more details on these *left self-distributive* rings, see [4].

3. THE MEDIAL RADICAL

Let $M(R) = < R[R,R]R > = \{\sum x_n(a_n b_n - b_n a_n)y_n : a_n, b_n, x_n, y_n, \in R\}$. Thus M is a function defined on the class of associative rings that assigns to every ring R a uniquely determined ideal $M(R)$ of R such that: (i) $h(M(R)) \subseteq M(h(R))$ for every ring homomorphism h; (ii) $M(R/M(R)) = 0$ (*i.e.*, $R/M(R)$ is medial); (iii) $M(X) \subseteq M(R)$ for any subring X of R. Hence M is a complete quasiradical in the sense of Maranda-Michler [18, p. 54]. Consequently, we will call $M(R)$ the *medial quasiradical* of R. Furthermore, a radical property, in the sense of Amitsur and Kurosh [18, p. 12] or [6, p. 3], can be defined via M by saying a ring X is a radical ring if $M(X) = X$. We define the *medial radical* of R, $\bar{M}(R)$, to be the sum of all radical ideals of R.

In this section we will discuss the structure of a ring R in terms of $M(R)$ and $\bar{M}(R)$ and show that R is a right essential extension of $M(R) + L(R)$, where $L(R) = l_R([R,R]R)$ is a medial ideal of R. Furthermore, we will show that the upper radical property determined by the class of commutative rings is the same as that determined by the class of rings satisfying a permutation identity and that this upper radical property is $\bar{M}$. Among ideals satisfying a permutation identity, $L(R)$ occupies a prominent position, for if K is an ideal which satisfies a permutation identity and $K \cap L(R) = 0$, then K is a right essential extension of a nilpotent ideal. Also, a finite sum of ideals each satisfying some permutation identity is an essential extension of a medial ideal. Hence a ring satisfying a permutation identity is "essentially" medial. Finally, an iterative procedure is developed which, in the case of a ring with D.C.C. on ideals, allows us to "shrink" $M(R)$ down to $\bar{M}(R)$ and to expand $L(R)$ to a larger medial ideal H of R such that $\bar{M}(R) + H$ is right essential in R.

The following basic facts and definitions will be used throughout this section.

(i) Let X be a right R-submodule of a right R-module Y. We write $X \leq'_R Y$ if, whenever $y \in Y$ and $y \notin X$, then there exists $s \in R$ such that $0 \neq ys \in X$. Observe that this condition implies that X is right essential in Y (*i.e.*, every nonzero R-submodule of Y has nonzero intersection with X).

(ii) $M(R) \subseteq < R,R >$.

(iii) If R has a right identity, then $L(R) = l_R([R,R])$ and $R[R,R] \subseteq M(R)$. Thus, if R has an identity, then $M(R) = < R,R >$.

(iv) If I is a left, right or two-sided ideal of R, then $M(I)$ is a left, right or two-sided ideal of R, respectively.

(v) Let I be an ideal of R. If $M(I) = I$, then $I^2 = I$; hence $[\bar{M}(R)]^2 = \bar{M}(R)$.

(vi) $M^1(R) = M(R)$, $M^2(R) = M(M(R))$, . . . , $M^{n+1}(R) = M^n(M(R))$, ...

(vii) $\bar{M}(R) \subseteq M^{n+1}(R) \subseteq [M^n(R)]^4 \subseteq M(R)$ for all positive integers n. Hence any nilpotent ring is $\bar{M}$-semisimple.

(viii) If there exists a positive integer k such that $M^{k+1}(R) = M^k(R)$, then $M^k(R) = \bar{M}(R)$.

(ix) If X is a subring of R such that $M(X) = X$, then $M(<X>) = <X>$. Thus $\bar{M}$ is a strong radical since it contains all one-sided $\bar{M}$-ideals of R[12, p. 49]. In fact, $\bar{M}(R)$ is the sum of all $\bar{M}$-right (left) ideals of R.

(x) If I is a minimal ideal, either $M(I) = I$, or $I^2 = 0$, or I is a field.

Proposition 3.1. The upper radical property determined by the class of commutative rings is the same as the upper radical property determined by the class *Perm*.

Proof. Let **C** be the class of all rings R such that every nonzero ideal of R can be mapped homomorphically onto some nonzero commutative ring. Let **P** be the class of all rings R such that every nonzero ideal of R can be mapped homomorphically onto some nonzero permutation identity ring. Clearly $\mathbf{C} \subseteq \mathbf{P}$. Let $R \in \mathbf{P}$ and I is a nonzero ideal of R. Then there exists a nonzero permutation identity ring X which is a homomorphic image of I. If $X \neq <X, X>$, then $X/<X,X>$ is a nonzero commutative ring, hence $R \in \mathbf{C}$. So assume $X = <X,X>$. Then X is nilpotent of index k. If $k = 2$, then again $R \in \mathbf{C}$. Suppose $X^2 \neq 0$. Then X/X^2 is a nonzero commutative ring. Thus in all cases $R \in \mathbf{C}$. Consequently, $\mathbf{C} = \mathbf{P}$. The result follows from [6, pp. 5-7].

Proposition 3.2. The upper radical property determined by the class of commutative rings is $\bar{M}$.

Proof. Let R be a ring. Clearly the homomorphic image of an $\bar{M}$-ring is an $\bar{M}$-ring and $\bar{M}(R)$ is an $\bar{M}$-ideal which contains all $\bar{M}$-ideals of R. Let $\bar{I}$ be an ideal of $R/\bar{M}(R)$ such that $M(\bar{I}) = \bar{I}$. Then there exists an ideal I of R such that $\bar{I} = I/M(R)$. Let $x \in I$ and $\bar{x}$ be the canonical image of x in $\bar{I}$. Since $\bar{x} \in M(\bar{I})$, then $x \in M(I) + \bar{M}(R)$. But $\bar{M}(R) \subseteq M(I)$. Hence $I = M(I)$, so $\bar{I} = 0$. Consequently, $\bar{M}(R/\bar{M}(R)) = 0$. Therefore, $\bar{M}$ is a radical property. From [6, pp. 5-7], it follows that $\bar{M}$ is the upper radical property determined by the class of commutative rings.

Proposition 3.3. Let I be an ideal of R. Then:

(i) $L(I) = l_I([I,I]I)$ is a medial ideal of R. Hence $L(I) < L(I), L(I) >_R L(I) = 0$.

(ii) $L(I) \leq'_I l_I(M(I))$ and $[l_I(M(I))]I \subseteq L(I)$. Also, if X is a right ideal of I such that $X \cap M(I) = 0$, then $X \subseteq L(I)$.

(iii) If K is a right ideal of I and $K \leq'_I L(I)$, then $M(I) + K \leq'_I$ I. In particular, $M(I) = 0$ if and only if $I = L(I)$.

(iv) If X is a right ideal of I, then $(X \cap M(I)) + (X \cap L(I)) \leq'_I X$.

Proof.

(i) Clearly $L(I) = \{x \in I : xabc = xbac \text{ for all } a, b, c \in I\}$ is a left ideal of R. Let $a, b, c \in I$, $x \in L(I)$, and $y \in R$. Then $xy(ab - ba)c = x(ya)bc - x(yb)ac = x(by)ac - x(ay)bc = xabyc - xbayc = x(ab - ba)yc = 0$. Thus $L(I)$ is a medial ideal of R.

(ii) Assume $x \in l_I(M(I))$ but $x \notin L(I)$. Then there exists $d \in [I, I]I$ such that $0 \neq xd$. Hence $0 = x(d(ab - ba)c) = xd(ab - ba)c$ for all $a, b, c \in I$. Thus $L(I) \leq'_I l_I(M(I))$. Similarly, $[l_I(M(I))]I \subseteq L(I)$ and $X \subseteq L(I)$.

(iii) Assume $x \in I$, but $x \notin M(I) + K$. If $x \in L(I)$, then there exists $y \in I$ such that $0 \neq xy \in K$. If $x \notin L(I)$, then there exists $a, b, c \in I$ such that $0 \neq x(ab - ba)c \in M(I)$. Therefore, $M(I) + K \leq'_I I$.

(iv) See proof of part (iii).

Example 3.4. Let $R = \begin{bmatrix} A & X \\ 0 & C \end{bmatrix}$, where A and C are rings and X is an $A-C$ bimodule.

(i) Let $A = X = C = F$ where F is a field. Then R is a subdirectly irreducible ring which has D.C.C. on ideals and is $\bar{M}$-semisimple. Since R has unity and is not commutative, R does not satisfy a permutation identity. Observe that $M(R) = \begin{bmatrix} 0 & F \\ 0 & 0 \end{bmatrix} \subseteq L(R) = \begin{bmatrix} 0 & F \\ 0 & F \end{bmatrix}$.

(ii) Let $A = X = D$ and $C = F$, where D is a noncommutative division ring and F is a subfield of D (e.g., center of D). Then R has D.C.C. on ideals, and $0 \neq \bar{M}(R) = M(R) \not\subseteq L(R)$. Observe that $\bar{M}(R) = \begin{bmatrix} D & D \\ 0 & 0 \end{bmatrix}$ and $L(R) = \begin{bmatrix} 0 & D \\ 0 & F \end{bmatrix}$. Note that $\bar{M}$ is not a hereditary radical since $\begin{bmatrix} 0 & D \\ 0 & 0 \end{bmatrix} \subseteq \bar{M}(R)$.

Proposition 3.5. Let K be a permutation identity ideal of R with Putcha-Yaqub number n.

(i) $K^{2n} < R, R > K^{2n} = (K < R, R >)^{4n} = 0$.

(ii) If $K \cap L(R) = 0$, then $K \cap M(R) \leq'_R K$ and $(K \cap M(R))^{5n} = 0$.

(iii) If K contains no nilpotent ideals of R, then $K \subseteq L(R)$.

(iv) If $M(R)$ contains no nonzero nilpotent ideal of R, then $L(R) = l_R(M(R))$. Hence $L(R)$ contains every permutation identity ideal of R.

(v) If P is a prime ideal such that R/P is not a commutative ring, then P contains every permutation identity ideal of R.

Proof. By [17], K^n is a medial ideal of R. Let $k_1, k_2, k_3, k_4 \in K^n$, and a, b $\in$ R. Consider $k_1(k_2 a)(bk_3)k_4 = k_1(bk_3)k_2(ak_4) = k_1(k_2 b)k_3(ak_4) = k_1 k_3(k_2 ba)k_4 = k_1 k_2 bak_3 k_4$. Hence, $k_1 k_2 (ab - ba)k_3 k_4 = 0$. Hence $K^{2n} < R, R > K^{2n} = (K < R, R >)^{4n} = 0$. The remainder of the proof is a consequence of part (i) and Proposition 3.3.

Corollary 3.6. Let K be a permutation identity ideal of R.

(i) $N(K)$ is an ideal of R. In particular, $N(L(R))$ is a nil ideal of R.

(ii) If $x \in N(R)$ and $y \in R$ such that $xy \in K$, then $xy \in N(K)$.

(iii) If X is an nil or nilpotent ideal of K, then $< X >_R$ is a nil or nilpotent ideal of R, respectively.

Proof. This result follows from a straightforward argument using the fact that there exists a positive integer m such that $K^m < R, R > K^m = 0$.

Proposition 3.7. Let I be an ideal of R and $I_1 = M(I) + L(I)$, $I_2 = M(I_1) + L(I_1)$, $\cdots$, $I_{n+1} = M(I_n) + L(I_n)$ for $n = 0, 1, 2, \cdots$, and $I = I_0$. Then:

(i) $L(I) \subseteq L(I_1) \subseteq \cdots \subseteq L(I_n) \subseteq \cdots$ is an ascending chain of medial ideals of R.

(ii) $I_{n+1} \leq'_{I_n} I_n$, hence $I_n \leq'_I I$.

(iii) $M^2(I_n) \subseteq M(I_{n+1}) \subseteq M^2(I_n) + [M(I_n)][L(I_n)]$.

(iv) $I_{n+1} = M^{n+1}(I) + L(I_n)$. In particular, if $M^{n+1}(I) = 0$, then $L(I_n) \leq'_I I$.

Proof.

(i) From Proposition 3.3, $L(I_n)$ is a medial ideal of R. Let $x \in L(I_n) \subseteq I_{n+1}$. Since $I_{n+1} \subseteq I_n$, then $x(ab - ba)c = 0$ for all a, b, $c \in I_{n+1}$. Hence $x \in L(I_{n+1})$.

(ii) Assume $x \in I_n$, but $x \notin I_{n+1}$. Then there exists a, b, $c \in I_n$ such that $0 \neq x(ab - ba)c \in M(I_n) \subseteq I_{n+1}$. Hence $I_{n+1} \leq'_{I_n} I_n$.

(iii) Clearly, $M^2(I_n) \subseteq M(I_{n+1})$. Let $v = (y_1 + a_1)[(y_2 + a_2)(y_3 + a_3) - (y_3 + a_3)(y_2 + a_2)](y_4 + a_4)$, where $y_i \in M(I_n)$ and $a_i \in L(I_n)$ for $i = 1, 2, 3, 4$. Since $[L(I_n)][M(I_n)] = 0$ (Proposition 3.3(ii)) and $a_1 \in L(I_n)$, then

$$v = y_1[y_2 y_3 + y_2 a_3 + a_2 a_3 - y_3 y_2 - y_3 a_2 - a_3 a_2](y_4 + a_4)$$

$$= y_1[y_2 y_3 - y_3 y_2]y_4 + y_1[y_2 y_3 + y_2 a_3 + a_2 a_3 - y_3 y_2 - y_3 a_2 - a_3 a_2]a_4$$

$$\in M^2(I_n) + [M(I_n)][L(I_n)] .$$

Hence

$$M^2(I_n) \subseteq M(I_{n+1}) \subseteq M^2(I_n) + [M(I_n)][L(I_n)] .$$

(iv) Clearly, $I_1 = M^1(I) + L(I_0)$. By part (iii), $M^2(I) \subseteq M(I_1) \subseteq M^2(I) + [M(I)][L(I)]$. But $[M(I)][L(I)] \subseteq L(I) \subseteq L(I_1)$, by part (i). Hence $I_2 = M^2(I) + L(I_1)$. Again, by part (iii), $M^2(I_1) \subseteq M(I_2) \subseteq M^2(I_1) + [M(I_1)][L(I_1)]$. Clearly, $M^3(I) \subseteq M^2(I_1) \subseteq M(M^2(I) + [M(I)][L(I)])$. If we take $v_1 \in M(M^2(I) + [M(I)][L(I)])$ and calculate in a manner similar to that for v in part (iii), we have $M^2(I_1) \subseteq M(M^2(I) + [M(I)][L(I)]) \subseteq M^3(I) + [M(I)][L(I)]$. Thus $M^3(I) \subseteq M(I_2) \subseteq M^3(I) + L(I_1)$. Hence $M^3(I) + L(I_2) \subseteq M(I_2) + L(I_2) = I_3 \subseteq M^3(I) + L(I_1) + L(I_2) = M^3(I) + L(I_2)$. Therefore, $I_3 = M^3(I) + L(I_2)$. Consequently, iterating on this process yields $I_{n+1} = M^{n+1}(I) + L(I_n)$ for $n = 0, 1, 2, \ldots$.

The following result shows that a permutation identity ring is "essentially" medial.

Corollary 3.8. If R is a permutation identity ring, then there is a medial ideal W such that $W \leq'_R R$.

Proof. This result follows from Proposition 3.7(iv) and Proposition 3.5.

From Example 3.4(i) we observe that $r_R([R,R]) = \begin{bmatrix} r & F \\ 0 & 0 \end{bmatrix}$ is a left permutable ideal. Now $R = l_R([R,R]) + r_R([R,R])$. Hence R is a sum of permutation identity ideals, but R does not satisfy a permutation identity. However, the next result shows that such a sum is "essentially" medial.

Corollary 3.9. Let $K = \sum_{i=1}^n K_i$, where the K_i are permutation identity ideals of R. Then there exists a medial ideal W of R such that $W \leq'_K K$.

Proof. Let $V = \sum(K_i \cap M(K))$. By Proposition 3.5, V is a nilpotent ideal of R. Let $P = V + L(K)$. Let $0 \neq \sum k_i \in K$, where $k_i \in K_i$. If $\sum k_i \notin L(K)$, then there exists a, $b, c \in K$ such that $0 \neq (\sum k_i)(ab - ba)c \in V$. Thus $P \leq'_K K$. Let $x \in M(P)$. Without loss of generality, assume $x = (v_1 + a_1)[(v_2 + a_2)(v_3 + a_3) - (v_3 + a_3)(v_2 + a_2)](v_4 + a_4)$, where $v_j \in V$ and $a_j \in L(K)$ for $j = 1, 2, 3, 4$. Since $[L(K)][M(K)] = 0$ (Proposition 3.3(ii)) and $a_1 \in L(K)$, then $x = v_1[v_2 v_3 + v_2 a_3 + a_2 a_3 - v_3 v_2 - v_3 a_2 - a_3 a_2](v_4 + a_4) = v_1[v_2 v_3 - v_3 v_2]v_4 + v_1[v_2 v_3 + v_2 a_3 + a_2 a_3 - v_3 v_2 - v_3 a_2 - a_3 a_2]a_4 \in M(V) + [M(K)][L(K)]$. Hence $M(P)$ is a nilpotent ideal of R. Thus there exists a positive integer n such that $M^n(P) = 0$. By Proposition 3.7(iv), there exists $W \leq'_K K$.

Corollary 3.10. Let R be a ring with D.C.C. on ideals and $R_1 = M(R) + L(R)$, $R_2 = M(R_1) + L(R_1), \ldots, R_{n+1} = M(R_n) + L(R_n)$ for $n = 0, 1, 2, \ldots$, and $R = R_0$. Then:

(i) There exists an integer n such that $\bar{M}(R) + L(R_n) \leq'_R R$.

(ii) If $\bar{M}(R) = 0$ and I is an ideal of R, then there exists a medial ideal K of R such that $K \leq'_I I$.

(iii) If $L(R_n) = 0$ and I is an ideal of R, then there exists an ideal Y or R such that $\bar{M}(I) + Y \leq'_I I$ and $Y^5 = 0$.

Proof.

(i) From Proposition 3.7 there exists n such that $R_n = R_{n+1} = \cdots$. Hence $M^{n+1}(R) = \bar{M}(R)$. Thus $\bar{M}(R) + L(R_n) \leq'_R R$.

(ii) This is a consequence of Proposition 3.7 and the fact that $\bar{M}(I) \subseteq \bar{M}(R)$.

(iii) From Propositions 3.3 and 3.7, there exists an integer n such that $I_n = I_{n+1} = \cdots$ and $I_{n+1} = M(I) + L(I_n) \leq'_I I$. Let $Y = L(I_n) \cap M(R)$. By Proposition 3.5, $Y^5 = 0$. By Proposition 3.7(i), $L(R) = 0$. From Proposition 3.3(iv), $Y \leq'_I L(I_n)$. Now Proposition 3.3(iii) and Proposition 3.7 yields $\bar{M}(I) + Y \leq'_I I$.

Proposition 3.11. Let Y be an ideal of R such that $M(Y) = Y$. Then $M(Y_{n \times n}) = Y_{n \times n}$ for any positive integer n. Consequently, $(\bar{M}(R))_{n \times n} \subseteq \bar{M}(R_{n \times n})$.

Proof. The result is a consequence of the following calculation which generalizes to the $n \times n$ case. Let $a_j, b_j, x_j, y_j \in Y$ for $j = 1, 2, 3, 4$. Then

$$
\begin{bmatrix} x_1(a_1b_1 - b_1a_1)y_1 & x_2(a_2b_2 - b_2a_2)y_2 \\ x_3(a_3b_3 - b_3a_3)y_3 & x_4(a_4b_4 - b_4a_4)y_4 \end{bmatrix}
$$

$$
= \begin{bmatrix} x_1 & x_2 \\ 0 & 0 \end{bmatrix} \left(\begin{bmatrix} a_1 & 0 \\ 0 & a_2 \end{bmatrix} \begin{bmatrix} b_1 & 0 \\ 0 & b_2 \end{bmatrix} - \begin{bmatrix} b_1 & 0 \\ 0 & b_2 \end{bmatrix} \begin{bmatrix} a_1 & 0 \\ 0 & a_2 \end{bmatrix} \right) \begin{bmatrix} y_1 & 0 \\ 0 & y_2 \end{bmatrix}
$$

$$
+ \begin{bmatrix} 0 & 0 \\ x_3 & x_4 \end{bmatrix} \left(\begin{bmatrix} a_3 & 0 \\ 0 & a_4 \end{bmatrix} \begin{bmatrix} b_3 & 0 \\ 0 & b_4 \end{bmatrix} - \begin{bmatrix} b_3 & 0 \\ 0 & b_4 \end{bmatrix} \begin{bmatrix} a_3 & 0 \\ 0 & a_4 \end{bmatrix} \right) \begin{bmatrix} y_3 & 0 \\ 0 & y_4 \end{bmatrix}.
$$

We note that a $\bar{M}$-radical ring can be a direct sum of medial left ideals. Let $R = \begin{bmatrix} F & F \\ F & F \end{bmatrix}$, where F is a field. Then R is a simple $\bar{M}$-radical ring. However, $R = \begin{bmatrix} F & 0 \\ F & 0 \end{bmatrix} \oplus \begin{bmatrix} 0 & F \\ 0 & F \end{bmatrix}$ is a direct sum of right permutable left ideals of R. Also observe that $(\bar{M}(F))_{2 \times 2} = 0 \neq \bar{M}(F_{2 \times 2}) = F_{2 \times 2}$. More generally, if R is any semiprime ring with unity, Proposition 2.15 shows that $R_{n \times n}$ $(n > 1)$ contains no medial ideals. If R also has D.C.C. on ideals, then by Corollary 3.10, $R_{n \times n}$ is "essentially" a $\bar{M}$-radical ring.

REFERENCES

1. G. Birkenmeier and H. Heatherly, Operation inducing systems, Algebra Universalis **24** (1987) 137-148.

2. G. Birkenmeier and H. Heatherly, Medial near-rings, Mh. Math. **107** (1989) 89-110.

3. G. Birkenmeier and H. Heatherly, Medial rings and an associated radical, Czech, Math. J., to appear.

4. G. Birkenmeier, H. Heatherly and T. Kepka, Rings with left self-distributive multiplication, submitted.

5. J.L. Chrislock, On medial semigroups, J. Algebra **12** (1969) 1-9.

6. N. Divinsky, *Rings and Radicals*, Univ. Toronto Press, 1965.

7. J. Fisher, Nil subrings with bounded indices of nilpotency, J. Algebra **19** (1971) 509-516.

8. I.N. Herstein, *Topics in Ring Theory*, Univ. Chicago Press, Chicago, 1965.

9. I.N. Herstein and L. Small, Nil rings satisfying certain chain conditions, Canad. J. Math. **16** (1964) 771-776.

10. J. Jezek and T. Kepka, *Medial Groupoids*, Rozpray CSAV, Rada mat. a prir. ved. 93/2(1983), Academia, Praha.

11. J. Jezek and T. Kepka, Permutable groupoids, Czech. Math. J. **34** (1984) 396-410.

12. A. Kertesz (ed.), *Rings, Modules and Radicals*, Colloquia Math. Soc. Janos Bolyai, 6, North-Holland, Amsterdam-London, 1973.

13. J. Levitzki, On nil subrings, Israel J. Math. **1** (1963) 215-216.

14. T. Nordahl, Residual finiteness in permutation varieties of semigroups, Semigroup Forum **31** (1985) 33-46.

15. T. Nordahl, On permutative semigroup algebras, Algebra Universalis **25** (1988) 322-333.

16. S. Pellegrini Manara, On medial near-rings, *Near-Rings and Near-Fields*, (Proc. Conf. Tubingen, 1985), North-Holland Math. Stud. **137** (1987), 199-209.

17. M. Putcha and A. Yaqub, Semigroups satisfying permutation identities, Semigroup Forum **3** (1971) 68-73.

18. F.A. Szasz, *Radicals of rings*, John Wiley and Sons, New York, 1981.

University of Southwest Louisiana
Department of Mathematics
Lafayette, Louisiana 70504-1010

QUANTUM GROUPS, FILTERED RINGS AND GELFAND-KIRILLOV DIMENSION

J.C. McConnell

ABSTRACT

It is shown here that, for a number of quantum groups, there exists a finite dimensional *'filtration'* for which the associated graded algebra has a simple form. It follows from this that Gelfand-Kirillov dimension behaves particularly well for these algebras.

1. Quantum Groups

1.1 The term *'quantum group'* is usually used, not only for the q-analogue of the coordinate ring of a semisimple algebraic group but also for the q-analogue of the universal enveloping algebra of a semisimple Lie algebra, but we will often refer to the latter as a *'quantum enveloping algebra'*. For background information on quantum groups see [S], and for any other unexplained terminology, see [McR].

1.2 A quantum group is a Hopf algebra but in this lecture we will only be concerned with the algebra structure and so the coalgebra structure will be totally ignored. Some examples of quantum groups now follow.

1.3 Quantum plane and quantum n-space. The simplest quantum group is the *'quantum plane'* $O_q(\mathbf{C}^2)$, whose algebra structure is $O_q(\mathbf{C}^2) = \mathbf{C}[x,y]$, $yx = q^{-2}xy$. Here $q \in \mathbf{C}$ with $q \neq 0$. Quantum n-space, $O_q(\mathbf{C}^n)$, is defined similarly, $O_q(\mathbf{C}^n) = \mathbf{C}[x_1, \dots , x_n]$ with $x_j x_i = q^{-2} x_i x_j$ for $i < j$.

1.4 Quantum $n \times n$ matrices and its relatives. We first define $O_q(M_2(\mathbf{C}))$. (This is actually a bialgebra rather than a Hopf algebra so strictly is not itself a quantum group.)
$$O_q(M_2(\mathbf{C})) = \mathbf{C}[a,b,c,d] = \mathbf{C}\left[\begin{smallmatrix} a & b \\ c & d \end{smallmatrix}\right] \,,$$

with the following relations:

Row, Column: $ba = q^{-2}ab$, $ca = q^{-2}ac$, $dc = q^{-2}cd$, $db = q^{-2}bd$,
Acute angle: $bc = cb$,
Obtuse angle: $ad - da = (q^2 - q^{-2})bc$.
(Note that when $q = 1$, this is the ordinary coordinate ring $O(M_2(\mathbf{C}))$, *i.e.*, the coordinate ring of the affine variety $\mathbf{C}^4$.) For a motivation for these relations, which is due to Kobyzev, see [M].

1.5 The algebra $O_q(M_n(\mathbf{C}))$ is defined as the algebra $\mathbf{C}[x_{ij}]$, $1 \leq i,j \leq n$, with the relations that for any 2×2 minor $\left(\begin{smallmatrix} a & b \\ c & d \end{smallmatrix}\right)$ of the $n \times n$ matrix (x_{ij}), the subalgebra $\mathbf{C}[a,b,c,d]$ of $O_q(M_n(\mathbf{C}))$ is isomorphic to $O_q(M_2(\mathbf{C}))$ in the obvious way, via $a \longrightarrow a$, etc. (Again $O_q(M_n(\mathbf{C}))$ is a bialgebra rather than a Hopf algebra.) Note that $O_q(M_2(\mathbf{C}))$ can be constructed via a sequence of skew-polynomial extensions (or Ore extensions) starting from $\mathbf{C}$, viz

$$\mathbf{C} \subset \mathbf{C}[a] \subset \mathbf{C}[a, b] \subset \mathbf{C}[a, b, c] \subset \mathbf{C}[a, b, c, d] .$$

Similarly $O_q(M_n(\mathbf{C}))$ can be constructed via a sequence of Ore extensions starting from the top left hand corner of (x_{ij}),

$$\mathbf{C} \subset \mathbf{C}[x_{11}] \subset \mathbf{C}[x_{11}, x_{12}] \subset \mathbf{C}[x_{11}, x_{12}, x_{21}] \subset \cdots .$$

1.6 **The quantum general linear group.** Before defining the q-versions of the general linear and special linear groups, we first discuss the algebra structure of the ordinary coordinate rings of these groups. As above, $O(M_n(\mathbf{C}))$ is the commutative polynomial algebra, $\mathbf{C}[x_{ij}]$, $1 \leq i, j \leq n$. Let α be the determinant of the $n \times n$ matrix (x_{ij}). Then, for $G = GL_n(\mathbf{C})$, the coordinate ring $O(G)$ is $\mathbf{C}[x_{ij}]_\alpha$, the localization of $\mathbf{C}[x_{ij}]$ with respect to the powers of α. For $G = SL_n(\mathbf{C})$, the coordinate ring $O(G)$ is $\mathbf{C}[x_{ij}]/(\alpha - 1)$. The quantum versions of these coordinate rings are constructed in a similar manner but starting from $O_q(M_n(\mathbf{C}))$ instead of $O(M_n(\mathbf{C}))$. In $O_q(M_n(\mathbf{C}))$ there is a central element β called the quantum determinant of (x_{ij}), see $[S]$. Then

$$O_q(SL_n(\mathbf{C})) = O_q(M_n(\mathbf{C}))/(\beta - 1)$$

and

$$O_q(GL_n(\mathbf{C})) = O_q(M_n(\mathbf{C}))_\beta .$$

All that we will need to know about these algebras is that they arise from $O_q(M_n(\mathbf{C}))$ as a homomorphic image and as a localization at the powers of a single central element, respectively.

2. Quantum Enveloping Algebras

2.1 For the remainder of this paper k denotes a field. If g is a Lie algebra over k, then $U(g)$ is a Hopf algebra. For g semisimple, there is a quantum enveloping algebra $U_q(g)$ which we now define. Recall that for $g = sl(2)$, $U(g) = \mathbf{C}[f, h, e]$ with the relations $[h, e] = 2e$, $[h, f] = -2f$ and $[e, f] = h$, (where $[a, b] = ab - ba$). Now $U_q(sl(2)) = \mathbf{C}[F, K, K^{-1}, E]$ with the relations $KE = q^2 EK$, $KF = q^{-2}FK$ and $EF - FE = (K^2 - K^{-2})/(q^2 - q^{-2})$. Note that $U_q(sl(2))$ can be constructed via a sequence of Ore extensions, $U_q(sl(2)) = \mathbf{C}[K, K^{-1}][E][F]$.

2.2 For a general semisimple Lie algebra g, the definition of $U_q(g)$ is considerably more complicated than for $g = sl(2)$. First, recall a number of standard results concerning $U(g)$, see e.g., $[H]$, $[K]$. Let g be finite-dimensional semisimple and $n = \operatorname{rank} g(= \text{dimension of a Cartan subalgebra.})$. To g there is associated the Cartan matrix $A(g) = (a_{ij}) \in M_n(\mathbf{Z})$ where $a_{ij} = 2(\alpha_i, \alpha_j)/(\alpha_i, \alpha_i)$, $\alpha_1, \ldots, \alpha_n$ being a set of simple roots. Note that the convention used here is that of $[K]$ and so the α_i-string of roots through α_j is

$$\alpha_j, \alpha_i + \alpha_j, \ldots, -a_{ij}\alpha_i + \alpha_j .$$

For $g = sl(n+1)$, the Cartan matrix has 2's on the diagonal, -1's on the subdiagonal and superdiagonal and zeros elsewhere. For the rank 2 algebras B_2 and G_2, the Cartan matrices are, respectively, $\left(\begin{smallmatrix} 2 & -2 \\ -1 & 2 \end{smallmatrix}\right)$ and $\left(\begin{smallmatrix} 2 & -3 \\ -2 & 2 \end{smallmatrix}\right)$. If R is a k-algebra and $x \in R$, then ad $x : R \longrightarrow R$ is $r \longmapsto xr - rx$. By a theorem of Serre, (see e.g., $[H]$), $U(g) = \mathbb{C}[f_i, h_i, e_i]$, $1 \le i \le n$, with the relations $[h_i, e_j] = a_{ij} e_j$, $[h_i, f_j] = -a_{ij} f_j$, $[e_i, f_j] = \delta_{ij} h_i$ and, for $i \ne j$,

$$(\text{ad } e_i)^{1-a_{ij}} (e_j) = 0 \ ,$$

and there is a similar ad-relation for the $f's$. Note that if $a_{ij} = 0$, then e_i and e_j commute.

In this presentation of $U(g)$, we only have generators e_α, f_α corresponding to α a simple root, and we pay for this reduced set of generators by having the complicated ad-relations between the $e's$ and between the $f's$.

2.3 We now define $U_q(g)$ for g finite dimensional semisimple. Recall, see e.g., $[K]$, p. 15, that if A is the Cartan matrix of g, then there is a diagonal matrix D, $D \in GL_n(\mathbb{Z})$, such that DA is a symmetric matrix. (For $g = sl(n+1)$, $D = Id$.) Then

$$U_q(g) = \mathbb{C}[F_i, K_i, K_i^{-1}, E_i] \ , \ 1 \le i \le n$$

with the relations: the K_i commute,

$$K_i E_j = q^{d_i a_{ij}} E_j K_i \ , \ K_i F_j = q^{-d_i a_{ij}} F_j K_i$$

and

$$[E_i, F_j] = \delta_{ij} (K_i^2 - K_i^{-2})/(q^{2d_i} - q^{-2d_i}) \ .$$

It remains to write down the analogues of the ad-relations in $U(g)$ and for this some extra notation is required. Let t be an indeterminate and let $m \in \mathbb{N}$. Set $[m]_t = (t - t^{-1}) \dots (t^m - t^{-m})$ and, for $m > n > 0$, set

$$\left[\begin{smallmatrix} m \\ n \end{smallmatrix}\right]_t = [m]_t / [n]_t [m-n]_t \ .$$

Then the analogue of the ad-relation for the e_i is

$$\sum_{\gamma=0}^{1-a_{ij}} (-1)^\gamma \left[\begin{smallmatrix} 1-a_{ij} \\ \gamma \end{smallmatrix}\right]_{q^{2d_i}} E_i^{1-a_{ij}-\gamma} E_j E_i^\gamma = 0 \ ,$$

and there is a similar relation for the F_i.

Examples. For $g = sl(n+1)$, these relations are

$$E_i^2 E_{i \pm 1} - (q^2 + q^{-2}) E_i E_{i \pm 1} E_i + E_{i \pm 1} E_i^2 = 0$$

and

$$[E_i, E_j] = 0 \text{ otherwise.}$$

Unlike $U_q(sl(2))$ it is not at all clear at this stage whether $U_q(g)$ is Noetherian.

3. Filtered Rings and Gelfand-Kirillov Dimension

3.1 Let R be a filtered algebra over a field k. (So $R = \cup R_n$, $n \geq 0$, where R_n is a k-subspace of R, $R_n \subset R_{n+1}$ for all n, and $R_i R_j \subset R_{i+j}$ for all i, j.) A filtration is finite-dimensional if $R_0 = k$ and $\dim_k R_n < \infty$ for all n. A filtration is standard if $R_n = R_1^n$ for all n. If R has a standard finite-dimensional filtration, then the Gelfand-Kirillov dimension $GK(R)$ is

$$GK(R) = \inf\{\gamma \mid \dim R_n \leq n^\gamma \text{ for } n >> 0\}.$$

More generally, if $R = \cup R_n$ is any finite dimensional filtration (*i.e.*, not necessarily standard) such that the associated graded ring gr R is a finitely generated k-algebra, then $GK(R)$ may be computed by the above formula applied to this filtration. If M is a filtered R-module, ($M = \cup M_n$, $n \geq 0$, with $R_j M_j \subset M_{i+j}$), then this filtration is called a good filtration if gr M is a finitely generated gr R module. If $M = \cup M_n$ is a good filtration of M, then $GK(M)$ can be defined by the similar formula to that used for $GK(R)$.

3.2 We now recall some of the special properties of $U(g)$, g finite dimensional.

3.3 The Nullstellensatz. The algebra $U = U(g)$ satisfies the Nullstellensatz over k, *i.e.*, endomorphisms of simple U-modules are algebraic over k and each prime ideal of U is an intersection of primitive ideals (see [McR, Chapter 9]).

3.4 Exactness of GK-dimension. Gelfand-Kirillov dimension is exact for $U = U(g)$, *i.e.*, if

$$0 \to L \to M \to N \to 0$$

is a short exact sequence of finitely generated U-modules, then $GK(M) = \max\{GK($ $GK(N)\}$ (see [McR], 8.4.8).

3.5 Multiplicities. If M is a U-module with a good filtration, then a multiplicity $e(M)$, $e(M) \in \mathbf{N}$, can be associated to M and $e(M)$ is independent of the choice of good filtration of M. Further if

$$0 \to L \to M \to N \to 0$$

is a short exact sequence of finitely generated U-modules, then one of the following holds (where $d(-) = GK(-) \in \mathbf{N}$)

(i) $d(L) < d(M) = d(N)$, $e(M) = e(N)$;
(ii) $d(L) = d(M) > d(N)$, $e(L) = e(M)$;
(iii) $d(L) = d(M) = d(N)$, $e(M) = e(L) + e(N)$;

(see [McR], 8.4.8).

3.6 The above properties are all consequences of the fact that $U = U(g)$ has a finite dimensional filtration such that gr U is a commutative finitely generated k-algebra.

3.7 We now consider conditions on the associated graded ring which are weaker than commutativity but which still yield the above properties.

Definition. A k-algebra R will be called semi-commutative if R has a set of k-algebra generators $x_1, \ldots, x_n$, $R = k[x_1, \ldots, x_n]$, such that, for each i, j, $x_i x_j = \lambda_{ij} x_j x_i$, where $0 \neq \lambda_{ij} \in k$.

3.8 Theorem. Let R be a k-algebra with a finite-dimensional filtration such that gr R is semi-commutative on homogeneous generators (having various degrees). Then

(i) R satisfies the Nullstellensatz over k;
(ii) GK dimension is exact for R;
(iii) to each finitely generated R-module M, there is a multiplicity $e(M) \in \mathbf{N}$ satisfying the properties given in 3.5.

Proof. The proof is similar to the proof for $U(g)$, but since we are not assuming that the finite dimensional filtration of R is a standard filtration, some details are more complicated.

(i) It is clear that gr R satisfies generic flatness over k([McR], 9.4.10), and hence R satisfies generic flatness over k([McR], 9.4.9). The result now follows as in [McR], 9.4.21.

(ii) It is clear that gr R is Noetherian and the result then follows as in [McR], 8.3.11.

(iii) Consider first the case when gr R is semi-commutative on homogeneous generators each having degree one. Then by [McP], to any finitely generated gr R-module N with a good filtration $\{N_n\}$, we can associate a polynomial $f \in \mathbf{Q}[x]$ such that dim $N_n = f(n)$ for $n >> 0$. Set $d(N) = $ degree $f = GK(N)$ and $e(N) = $ leading coefficient of $f/d(N)!$ Then $e(N) \in \mathbf{N}$. If M is a finitely generated R-module with a good filtration set $d(M) = d(gr\ M)$ and $e(M) = e(gr\ M)$. The integers d and e behave as required on short exact sequences.

Consider now the case when gr R is semi-commutative on homogeneous generators having various degrees. Let h be the least common multiple of these degrees. Let L be a filtered gr R-module with a good filtration $\{L_n\}$. Then, as in [Lo], Chap. 3, Sect. 4.3, there is a family of polynomials $\{f_j | 0 \leq j \leq h\}$ such that, for $n >> 0$, dim $L_n = f_j(n)$ where $n \equiv j$ mod h, and the f_j all have the same degree and the same leading coefficient. So again one can define $d(M)$ and $e(M)$ for a finitely generated R-module M and these integers behave as required on short exact sequences.

4. Filtrations on quantum groups

4.1 In this section it will be shown that a number of the quantum groups considered in Sections 1 and 2 do have a finite dimensional filtration similar to that described in Theorem 3.8.

4.2 A filtration is defined on $O_q(M_n(\mathbf{C}))$ by assigning a degree to each of the generators x_{ij}. The degrees of the x_{ij} are given by the numbers in Pascal's triangle which is listed, like the x_{ij}, as

$$1\ 1\ 1\ 1\ ...$$
$$1\ 2\ 3\ 4\ ...$$
$$1\ 3\ 6\ ...$$
$$1\ 4\ ...$$
$$...$$
$$...$$

(That the degrees of the x_{ij} can be assigned like this can be seen from the construction of $O_q(M_n(\mathbf{C}))$ from $\mathbf{C}$ by a sequence of Ore extensions starting from the top left-hand corner of (x_{ij}).)

4.3 Theorem. The algebra $R = O_q(M_n(\mathbf{C}))$ (or $O_q(SL_n(\mathbf{C}))$ or $O_q(GL_n(\mathbf{C}))$, respectively) has a finite-dimensional filtration for which the associated graded ring is semi-commutative on homogeneous generators.

Proof. If $\begin{pmatrix} a & b \\ c & d \end{pmatrix}$ is a 2×2 minor of (x_{ij}), it is clear from the choice of degrees that

$$\deg a + \deg d > \deg b + \deg c\ ;$$

So in the associated graded ring $\bar{a}\bar{d} - \bar{d}\bar{a} = 0$ and hence gr R is semi-commutative.

4.4 The next step is to find a similar filtration for $U_q(g)$ but this is more difficult. For the remainder of this paper, we concentrate on the case $g = sl(n+1)$ where we are able to use results of Yamane [Y]. Following [Y], denote E_i by $e_{i,i+1}$ and F_i by $f_{i,i+1}$, and then define e_{ij}, f_{ij} for $1 \leq i < j \leq n$ inductively by

$$e_{ij} = q\ e_{i,j-1}e_{j-1,j} - q^{-1}e_{j-1,j}e_{i,j-1}$$

and similarly for the f_{ij}. (Intuitively, where the E_i and F_i correspond to the simple roots, elements corresponding to the other roots have now been constructed.) So

$$U_q(g) = \mathbf{C}[K_i, K_i^{-1}, e_{ij}, f_{ij}]\ ,\quad 1 \leq i < j \leq n\ .$$

The next step is to write down the relation between these generators. Order the $(i,j) \in \mathbf{N} \times \mathbf{N}$ lexicographically, i.e., $(i,j) < (i',j')$ if $i < i'$ or $i = i'$ and $j < j'$. There are six possible ways to have $(i,j) < (m,n)$:

I. $i = m < j < n$,
II. $i < m < n < j$,
III. $i < m < j = n$,
IV. $i < m < j < n$,
V. $i < j = m < n$,
VI. $i < j < m < n$.

The relations among the e_{ij} are:

for $(i,j) < (m,n)$ in the lexicographic ordering,

I or III $\quad e_{ij}e_{mn} = q^2 e_{mn}e_{ij}$,

II or VI $e_{ij}e_{mn} = e_{mn}e_{ij}$,

IV $[e_{ij}, e_{mn}] = (q^2 - q^{-2})e_{in}e_{mj}$,

V $q\,e_{im}e_{mn} - q^{-1}e_{mn}e_{im} = e_{in}$.

There are similar relations among the f_{ij}. The relations between the e_{ij} and f_{ij} are:

for $(i,j) < (m,n)$ in the lexicographic ordering

I $[e_{ij}, f_{mn}] = (-1)^{j-i+1}q\,f_{jn}K_i^2 K_{i+1}^2 \cdots K_{j-1}^2$,

III $[e_{ij}, f_{mn}] = (-1)^{n-m+1}q\,K_m^2 K_{M+1}^2 \cdots K_{j-1}^2 e_{im}$,

IV $[e_{ij}, f_{mn}] = (-1)^{j-m+1}(q^4 - 1)f_{jn}K_m^2 \cdots K_{j-1}^2 e_{im}$,

II, V, VI $[e_{ij}, f_{mn}] = 0$,

I $[e_{mn}, f_{ij}] = (-1)^{j-i}q^{-1}K_i^{-2} \cdots K_{j-1}^{-2} e_{jn}$,

III $[e_{mn}, f_{ij}] = (-1)^{n-m}q^{-1}f_{im}K_m^{-2} \cdots K_{n-1}^{-2}$,

IV $[e_{mn}, f_{ij}] = (-1)^{j-m}(1 - q^{-4})f_{im}K_m^{-2} \cdots K_{j-1}^{-2} e_{jn}$

II, V, VI $[e_{mn}, f_{ij}] = 0$.

Finally,

$$[e_{ij}, f_{ij}] = (-1)^{j-i+1}(K_i^2 \cdots K_{j-1}^2 - K_i^{-2} \cdots K_{j-1}^{-2})/(q^2 - q^{-2}) .$$

4.5 Yamane proves a Poincaré-Birkhoff-Witt theorem for $U_q(sl(n+1))$ as follows. Call the word

$$f_{k_1 m_1} \cdots f_{k_s m_s} K_1^{l_1} \cdots K_n^{l_n} e_{i_1 j_1} \cdots e_{i_t j_t}$$

a standard word if

$$(k_1, m_1) \leq \cdots \leq (k_s, m_s) \quad \text{and} \quad (i_1, j_1) \leq \cdots \leq (i_t, j_t) .$$

Then the set of standard words is a k-basis for $U_q(sl(n+1))$ (provided $q^8 \neq 1$). Yamane also defines a "filtration" (strictly a bifiltration) of $R = U_q(sl(n+1))$, so $R = \cup R_{m,n}$, $m, n \in \mathbf{N}$. The elements of $\mathbf{N} \times \mathbf{N}$ are ordered lexicographically and are called bidegrees. An element K_i is given bidegree (0,0) and the elements e_{ij} and f_{ij} are both given bidegree $(j - i, (j - i)i)$. Finally the bidegree of a standard word is the sum of the bidegrees of its terms. Then $R_{m,n}$ is the linear span of the standard words having bidegree at most (m,n). Let gr R denote the associated bigraded ring of R, gr $R = \oplus_{(m,n)}\bar{R}_{m,n}$. (Note that for a fixed m, there are only finitely many n for which $\bar{R}_{m,n} \neq 0$.)

4.6 **Theorem (Yamane).** gr R is a semi-commutative algebra on homogeneous generators (viz. the images of the K_i, K_i^{-1}, e_{ij} and f_{ij}).

4.7 In [McS], Corollary 1.7, it was shown that if S is a filtered algebra such that gr S is a finitely generated commutative algebra, then S has a finite dimensional filtration for which gr S is a finitely generated commutative algebra. This argument can be generalized to replace '$commutative$' by '$semi - commutative$' and replace '$filtration$' by '$bi - filtration$'. Thus

4.8 Theorem. Let $R = U_q(sl(n + 1))$ with $q^8 \neq 1$. Then R has a finite dimensional bifiltration, $R = \cup R_{m,n}$, such that gr $R = \oplus \bar{R}_{m,n}$ is semi-commutative on homogeneous generators.

We now deduce three consequences of this theorem.

(α) Since gr R is semi-commutative, gr R satisfies generic flatness over k and this property is readily pulled back to R. Hence we have

4.9 Corollary. R satisfies the Nullstellensatz over k.

(β) Since gr R is Noetherian, we have, by an extension of [McR] 8.3.11,

4.10 Corollary. Gelfand Kirillov dimension is exact for R.

(γ) Let M be an R-module with a good bifiltration $\{M_{m,n}\}$ and let gr M be the associated bigraded module over gr R. Now we can regard gr R as a graded ring on homogeneous semi-commutative generators by setting:

$$m^{th} \text{ homogeneous component of gr } R = \underset{n}{\oplus} \bar{R}_{m,n} ,$$

and, in a similar way, we can regard gr M as a graded module over this graded ring. So by applying Theorem 3.8 to gr R, we can associate to gr M a multiplicity $e(\text{gr } M)$. Now any two good bifiltrations of M will have bounded difference (compare [McR] 8.6.12) and hence $e(\text{gr } M)$ will be independent of the choice of the good bifiltration of M. Set $e(M) = e(\text{gr } M)$. Then

4.11 Corollary. On short exact sequences of R-modules the integers $d(-) = GK(-)$ and $e(-)$ behave as described in 3.5.

4.12 PBW theorems for some of the other $U_q(g)$ can be found in [Lu] and [T]. We hope to apply the $sl(n + 1)$ methods to these other $U_q(g)$ in a subsequent publication.

References

[H] Humphreys, J.E. *Introduction to Lie algebras and representation theory.* Graduate Texts No. 9. Springer Verlag, New York-Berlin, 1972.

[K] Kac, V. *Infinite dimensional Lie algebras,* 2nd ed. Cambridge University Press, Cambridge-New York, 1985.

[Lo] Lorenz, M. Gelfand-Kirillov dimension and Poincaré series. Cuadernos de Algebra No. 7. Universidad de Granada, 1988, pp. 68.

[Lu] Lusztig, G. Canonical bases arising from quantized enveloping algebras. Preprint, M.I.T., 1989.

[McP] McConnell, J.C. and Pettit, J.J. Crossed products and multiplicative analogues of Weyl algebras. J. London Math. Soc. **38** (1988) 47-55.

[McR] McConnell, J.C. and Robson, J.C. *Noncommutative Noetherian rings.* J. Wiley and Sons, Chichester-New York, 1987.

[McS] McConnell, J.C. and Stafford, J.T. Gelfand-Kirillov dimension and associated graded modules. J. Algebra **125** (1989) 197-214.

[M] Manin, Yu. I. Quantum group and non-commutative geometry. Les. Publ. du Centre de Récherches Math. Université de Montreal, 1988.

[S] Smith, S.P. Quantum groups: An introduction and survey for ring theorists. Preprint, University of Washington, Seattle, 1989.

[T] Takeuchi, M. The q-bracket product and the P.B.W. theorem for quantum enveloping algebras of classical types (A_n), (B_n), (C_n) and (D_n). Preprint, University of Tsukuba, 1989.

[Y] Yamane, H. A P.B.W. theorem for quantized universal enveloping algebras of type A_n. Publ. R.I.M.S. Kyoto **25** (1989) 503-520.

School of Mathematics
University of Leeds
Leeds, LS2, 9JT, ENGLAND

ORE LOCALIZATION IN THE FIRST WEYL ALGEBRA

Bruno J. Müller and Ying-Lan Zhang

1. INTRODUCTION

We prove that every "at most linear" localization, of the first Weyl algebra, is an Ore localization.

The first Weyl algebra, A, over the complex numbers, C, is the algebra of (ordinary) differential operators with polynomial coefficients. It is a simple Noetherian domain, of Krull and global dimension one. It can be presented as $A = C\{q, p\}$, subject to $pq = qp + 1$.

The hereditary torsion theories, in the category Amod of left A-modules, are classified by the sets of simple modules, and the latter were classified by Block [1]. In particular, every simple module has a "degree" (which one can think of as the order of attached differential operators). Some of the torsion theories are Ore localizations, for instance, the Goldie torsion theory (which corresponds to the set of all simple modules), and the next largest torsion theories (which correspond to the sets of all but one simple module, cf. Goodearl [4], 3.2). More significantly, some "small" torsion theories (corresponding to a single simple module S) are also Ore; this has been long known if S has degree zero and is proved here if S has degree one.

This progress was made possible by the fortunate fact that linear differential operators can be controlled to some extend: we know the similarity relation explicitly, we have a generous supply of "dense" members, and we have an effective description of the elements of a linear simple module.

All of this becomes much harder in degree two. A subsequent paper will contain a number of technical observations in this direction–though we have not been able yet to achieve decisive results on localization. Much to our initial surprise, there is no example known of a torsion theory in Amod which is not an Ore localization.

In the present paper, we have tried to avoid much of these technicalities and have given ad hoc arguments which are sometimes special to the linear case.

Precise definitions will be found in the body of the paper; background information can be obtained from McConnell and Robson [6], Stenström [7] and Golan [3]. The letters A, B and K are reserved for the first Weyl algebra, its localization at the set of all nonzero polynomials, and its quotient division ring, respectively. Modules are left A-modules unless specified otherwise. EX is the injective hull of the A-module X. The letters S and γ always stand for a simple A-module and an isomorphism-closed class of simple A-modules, respectively.

2. TORSION THEORIES

We list a few results about arbitrary (hereditary) torsion theories in Amod, most of which depend only on the fact that A is an HNP ring, and in particular has Krull dimension one:

Every torsion theory τ in Amod is perfect. If $\tau \neq Amod$, then τ is faithful, *i.e.,* A embeds into the quotient ring. Every such quotient ring is an overring of A, *i.e.,* an intermediate ring between A and K. [Conversely every overring is the quotient ring of a unique torsion theory. Every quotient ring is a simple Noetherian domain.]

A faithful torsion theory in Amod is determined by the set γ of simple torsion modules (use [7], VI.2.5, 3.3). This establishes a one-to-one correspondence between faithful torsion theories τ in Amod, and isomorphism-closed classes γ of simple left A-modules; we shall write $\tau = {}_\gamma\tau$ if τ corresponds to γ.

There is a duality between left and right A-modules of finite length, implemented by the functor $(-)^* = \hom_A(-, K/A) \simeq Ext_A^1(-, A)$. In particular, S^* is a simple right A-module, for any simple left A-module S. We use $\gamma*$ for the isomorphism-closure of the set $\{S^* : S \in \gamma\}$. Thus we obtain a one-to-one correspondence between the torsion theories in Amod and mod A, via ${}_\gamma\tau \longleftrightarrow \tau_{\gamma*}$. [It can be shown that the quotient rings of ${}_\gamma\tau$ and $\tau_{\gamma*}$ coincide.]

Definition and Lemma 1. An element $x \in A$ will be called γ-*dense* if it has the following equivalent properties:

(1) $A/Ax \in {}_\gamma\tau$;

(2) x is invertible in the quotient ring Q of ${}_\gamma\tau$;

(3) x acts regularly on ES' of all simple $S' \notin \gamma$;

(4) $\hom_A(A/Ax, ES') = 0$ for all simple $S' \notin \gamma$;

(5) $x \neq 0$, and all composition factors of A/Ax belong to γ.

Proof.

(1) $\Rightarrow$ (2): As ${}_\gamma\tau$ is perfect, the quotient functor is $Q \otimes_A -$, where Q is the quotient ring. Then $A/Ax \in {}_\gamma\tau$ implies $0 = Q \otimes (A/Ax) \simeq Q/Qx$; consequently $1 = qx$ for some $q \in Q$. As Q is a domain, q is a two-sided inverse for x.

(2) $\Rightarrow$ (3): If $S' \notin \gamma$, then ES' is ${}_\gamma\tau$-torsionfree and injective, hence a Q-module. As x is invertible in Q, it acts regularly on ES'.

(3) $\Rightarrow$ (4): If $f \in \hom_A(A/Ax, ES')$, then $xf(\bar{1}) = f(\bar{x}) = 0$, hence $f(\bar{1}) = 0$ by assumption, hence $f = 0$.

(4) $\Rightarrow$ (5): $x \neq 0$ holds since $\hom_A(A, ES') \neq 0$. Then A/Ax has finite length. If there was a composition factor $S' \notin \gamma$, one would obviously obtain a nonzero homomorphism $A/Ax \to ES'$.

(5) $\Rightarrow$ (1): This is obvious since $\gamma \subseteq {}_\gamma\tau$.

We shall denote the multiplicatively closed set of all γ-dense elements of A by $\Delta(\gamma)$. It can be easily seen (cf. [4], 2.3) that x acts regularly on the left module ES iff it acts regularly on the right module ES*. Consequently $\Delta(\gamma) = \Delta(\gamma*)$.

3. THE ORE CONDITION

Any multiplicative subset Σ of any ring R determines a torsion theory $_\Sigma\tau$ in Rmod: a module is torsion iff every element is annihilated by a member of Σ.

We list two easy general facts:

Lemma 2. A multiplicative subset Σ of a left Noetherian ring R is left Ore iff Σ acts regularly on every (indecomposable injective) Σ-torsionfree left R-module.

Lemma 3. If the multiplicative subsets Σ_i of the arbitrary ring R are left Ore $(i \in I)$, then the multiplicative set Σ generated by all Σ_i is left Ore, and $_\Sigma\tau = V_{i \in I}$ $_{\Sigma_i}\tau$.

Proof. For the non-trivial inclusion, $_\Sigma\tau \subseteq V_{\Sigma_i}\tau$, use ([7], VI.2.5, 3.3) and the fact that the Σ_i are left Ore.

Returning to the Weyl algebra A, we call a torsion theory $\tau = {_\gamma\tau}$, as well as the corresponding class γ of simple A-modules, *Ore* if τ is determined by a left Ore set.

As A is "highly non-commutative", one would expect to find many torsion theories which are not Ore. However, no such example appears to be known. On the other hand, one knows many Ore torsion theories, and we shall add to that list.

Lemma 4. A torsion theory in Amod is Ore, iff $_\gamma\tau = {_{\Delta(\gamma)}\tau}$. If so, then $\Delta(\gamma)$ is a left Ore set.

Proof. ($\Rightarrow$) : We are given $\tau =_\Sigma \tau$, for some left Ore set Σ. If $x \in \Sigma$, then $A/Ax \in {_\Sigma\tau}$, since Σ is left Ore. Lemma 1 shows $x \in \Delta(\gamma)$; hence $\Sigma \subseteq \Delta(\gamma)$ and $_\gamma\tau =_\Sigma \tau \subseteq {_{\Delta(\gamma)}\tau}$. The reverse inclusion, $_{\Delta(\gamma)}\tau \subseteq {_\gamma\tau}$, is always true; cf. Lemma 1.

($\Leftarrow$) : We show that $\Delta(\gamma)$ is left Ore, by verifying the assumptions of Lemma 2. As A has Krull dimension one, the indecomposable injective A-modules are K and ES for the simple A-modules S. Obviously $\Delta(\gamma)$ acts regularly on K, and also on ES if $S \notin \gamma$. If $S \in \gamma$, then $S \in {_\gamma\tau} = {_{\Delta(\gamma)}\tau}$, hence ES is not $\Delta(\gamma)$-torsionfree, and nothing has to be checked.

Lemma 4 shows that $\Delta(\gamma)$ is the natural candidate, for a left Ore set to determine $_\gamma\tau$. We also obtain the following criterion, to check whether this is indeed the case:

Corollary 5. $\Delta(\gamma)$ is left Ore and determines $_\gamma\tau$, iff for every $s \in S \in \gamma$ there exists $x \in \Delta(\gamma)$ with $xs = 0$.

Proof. The criterion expresses the condition $\gamma \subseteq {_{\Delta(\gamma)}\tau}$, which is equivalent to $_\gamma\tau = {_{\Delta(\gamma)}\tau}$.

4. SIMPLE MODULES

The irreducible representations of the Weyl algebra were determined by Block [1]. His classification employs the overring $B = C(q)\{p\}$, the algebra of differential operators with rational coefficients, which is a left and right principal ideal domain. B is the localization of A at the two-sided Ore set of all nonzero polynomials in q.

We call a simple A-module S B-torsion [B-torsionfree] if it is torsion [torsionfree] for this localization, i.e., if $B \otimes_A S = 0 [\neq 0]$. Block's classification runs as follows: The B-torsion S are precisely (up to isomorphism) the $A/A(q - \rho)$, $\rho \in \mathbb{C}$. The B-torsionfree S are in one-to-one correspondence with the simple B-modules T, via $S \cong$ socle $_A T$.

The simple B-modules are of the form $T \cong B/Bb$, where b is any irreducible element of B. If $B/Ba \cong B/Bb$, then a and b are called similar, $a \sim b$. The similarity class of b is denoted by $[b]$, and also by $[S]$, where $S \cong$ socle $_A B/Bb$ is the corresponding simple A-module.

The most immediate invariant of $[b]$ is the p-degree of b (written as a polynomial in p with coefficients in $\mathbb{C}(q)$; this degree equals the $\mathbb{C}(q)$-vector space dimension of B/Bb). This degree is at least one. We transfer it to the corresponding B-torsionfree simple A-module S, and speak of the "degree of S". In particular we call S linear [quadratic] if this degree is one [two]. We extend the degree to B-torsion simple A-modules S, by assigning them the degree zero.

Remarks. It is easy to see that, for irreducible a, $b \in B$, $a \sim b$ holds iff there are $u, v \in B$, of strictly lower degree, such that $ua = bv$. An effective internal description of $a \sim b$ is available only in the linear case: here $a \sim b$ for monic a, b iff $a - b = f'/f$, a "logarithmic derivative", for some $f \in \mathbb{C}(q)$.

To illustrate how much more complicated this matter becomes in the quadratic case, we describe the elements $a \in B$ which are similar to the irreducible $b = p^2 - q^3$. They are precisely the $a = \chi(\alpha p^2 - \alpha' p + \beta)$, where $\alpha = \varphi^2 + \varphi \psi' - \varphi' \psi - q^3 \psi^2$, $\alpha' = d A/dq$ and $\beta = (\varphi' + q^3 \psi)(2\varphi' + \psi'' + q^3 \psi) - (\varphi + \psi')(\varphi'' + 3q^2 \psi + q^3 \varphi + 2q^3 \psi')$ and $\chi, \varphi, \psi \in \mathbb{C}(q)$, χ nonzero, and φ and ψ not both zero.

[This unwieldy description is not entirely without use: with the help of some of the technical results of the subsequent paper, one can deduce that $b = p^2 - q^3$ is the only S-dense member of its similarity class $[b] = [S]$. This scarcity of S-dense elements suggests that this particular simple module S might not be Ore.]

5. ORE LOCALIZATION

The torsion theories $_S \tau$ corresponding to the isomorphism class of a single simple A-module S (i.e., the atoms in the lattice of faithful torsion theories) are crucial for the investigation of the Ore condition. This is so because, due to Lemma 3, γ is Ore if all $S \in \gamma$ are Ore. Thus, if there exists an example of a torsion theory which is not Ore, then there is one among the $_S \tau$.

It is well known that all simple A-modules of degree zero are Ore. [To show that, for instance, A/Aq is Ore, one verifies the left Ore condition for the set $\{q^n : n \in \mathbb{N}\}$ directly: commuting p and q introduces a derivative; hence one has $q^{n+m} p^m = \sum_{i=0}^m c_i p^i q^{n+i} = (\sum_{i=0}^m c_i p^i q^i) q^n$ for suitable $c_i \in \mathbb{C}$. Consequently if $x \in A$ has p-degree m, one obtains $q^{n+m} x = y q^n$ for some $y \in A$.]

Our aim is to extend this result to linear simple modules. Throughout the rest of this section, and without further mention, S will denote a (usually linear) B-torsionfree simple A-module, and a an element of A in the corresponding similarity class, i.e., $a \in [S] \cap A$. If S is linear, we write $a = \alpha p - \beta$ with $\alpha \neq 0$, $\beta \in C[q]$. We first determine when a is S-dense.

Lemma 6. $a = \alpha p - \beta$ is S-dense, iff $(\alpha, \beta) = 1$ and β/α has no integer residue at any simple pole.

Proof. ($\Leftarrow$) : Block's indicial polynomial (cf. [1], Section 3.2), of a at $\rho \in C$, has in our special case the following simple description: $\theta_{a,\rho}(\xi) = \alpha(\rho)\xi$ if $\alpha(\rho) \neq 0$, $\theta_{a,\rho}(\xi) = \alpha'(\rho)\xi - \beta(\rho)$ if $\alpha(\rho) = 0$. It has the root $\xi = 0$ if α/β has no pole, the root $\xi = \beta(\rho)/\alpha'(\rho) = \mathrm{Res}_\rho \beta/\alpha \neq 0$ if β/α has a simple pole, and no root if β/α has a multiple pole, at ρ.

Thus our condition says that $\theta_{a,\rho}$ has no integer root at simple poles of β/α. Consequently $\theta_{a,\rho}$ has no nonzero integer root at any $\rho \in C$. In particular a is "preserving" ([1], Section 3.4) and, therefore, $S \simeq A/(A \cap Ba)$([1], Theorem 4.3). We proceed to show $A \cap Ba = Aa$.

Suppose not. Then $(A \cap Ba)/Aa$ is semisimple (cf. [5], 5.7) and B-torsion (since it is the kernel of the quotient map $A/Aa \to B/Ba$). Thus there is $0 \neq \bar{x} \in (A \cap Ba)/Aa$ with $(q - \rho)\bar{x} = 0$ for some $\rho \in C$. We deduce $(q - \rho)x = ra$ where $r \in A - (q - \rho)A$, hence $ra \equiv 0$ (modulo $(q - \rho)A$). We may assume $r = \sum_{i=0}^n r_i p^i$ with $r_i \in C$ and $r_n \neq 0$, $n \geq 0$. Thus $ra = (\sum_{i=0}^n r_i p^i)(\alpha p - \beta) = r_n \alpha(\rho)p^{n+1} + $ (lower terms in p) or $= r_n(\alpha'(\rho)n - \beta(\rho))p^n + $ (lower terms in p), according to whether $\alpha(\rho) \neq 0$ or $= 0$. By our assumption, the highest coefficient is always nonzero, and this is a contradiction. We conclude $A \cap Ba = Aa$, hence $S \simeq A/Aa$, and consequently $a \in \Delta(S)$ as claimed.

($\Rightarrow$) : This time we are given $S \simeq A/Aa$. Obviously $(\alpha, \beta) = 1$ follows. Suppose $\alpha(\rho) = 0$ and $\alpha'(\rho)n - \beta(\rho) = 0$ for some $0 \neq n \in \mathbf{Z}$. In B we compute $a(q - \rho)^n = (q - \rho)^{n+1}b$ where $b = \frac{\alpha}{q-\rho}p + \frac{1}{q-\rho}\left(\frac{n\alpha}{q-\rho} - \beta\right) \in C[q]$, by the supposition. If $n > 0$, then $A/Aa(q - \rho)^n$ has precisely n composition factors isomorphic to $A/A(q - \rho)$, but $A/A(q - \rho)^{n+1}b$ has at least $n + 1$ such composition factors; a contradiction. If $n < 0$, then $(q - \rho)^{-n-1}a = b(q - \rho)^{-n}$ leads to a similar contradiction.

Theorem 7. Every linear simple A-module is Ore.

Proof. Lemma 6 and ([1], 7.1) show that every linear similarity class $[S]$ contains an S-dense member, $a = \alpha p - \beta$, and that S is isomorphic to $C[q, \alpha^{-1}]$, with p acting as $d/dq + \beta/\alpha$. Using this presentation of S, we shall verify the criterion of Corollary 5.

Define, for all $k \in \mathbf{Z}$, $a_k = a - k\alpha'$. Then $\alpha a_k = \alpha(\alpha p - \beta - k\alpha') = \alpha(p\alpha - \alpha' - \beta - k\alpha') = a_{k+1}\alpha$. From $A/A\alpha a_k = A/Aa_{k+1}\alpha$, it follows that A/Aa_k and A/Aa_{k+1} have the same composition factors. But $A/Aa_0 = A/Aa \cong S$ is simple and, therefore, $A/Aa_k \cong S$ for all k. We conclude that all a_k are similar to a, and are S-dense.

In $C[q, \alpha^{-1}]$ we compute $a_k \cdots a_0 q^k = a_k \cdots a_1 (\alpha p - \beta) q^k = a_k \cdots a_1 (\alpha (d/dq + \beta/\alpha) - \beta) q^k = a_k \cdots a_1 \alpha k q^{k-1} \cdots a_0 q^{k-1} = \cdots = \alpha^k k! a q^0 = 0$. Consequently $a_k \cdots a_0$ annihilates every polynomial $f(q)$ of degree $\leq k$.

Now consider any element $s = \alpha^{-m} f \in C[q, \alpha^{-1}]$, where f has degree k. Define $x = a_{k-m} \cdots a_{-m} \in \Delta(S)$. Then $\alpha^m x s = \alpha^m a_{k-m} \cdots a_{-m} \alpha^{-m} f = a_k \cdots a_0 \alpha^m \alpha^{-m} f = 0$. As S is B-torsionfree, the factor α^m can be cancelled, and we obtain $xs = 0$ as required.

Corollary 8. Let γ be a class of simple left A-modules of degree ≤ 1. Then $\Delta(\gamma)$ is a two-sided Ore set.

Proof. As indicated at the beginning of the section, Lemma 3 implies that $\Delta(\gamma)$ is a left Ore set. Obviously duality preserves degree, so $\gamma*$ consists of simple right A-modules of degree ≤ 1. Consequently the right analog of Theorem 7 shows that $\Delta(\gamma) = \Delta(\gamma*)$ is right Ore.

We have no "genuine" example, of a simple A-module of degree greater than one, which is Ore. One can, however, easily produce "fake" examples of this kind, by observing that the Ore condition is preserved under automorphisms of A, and linear simple modules can be carried to non-linear ones.

For instance, the quadratic element $a = p^2 - q^2$ is irreducible in B and is carried by the automorphism $p \to (q + p)/\sqrt{2}$, $q \to (q - p)/\sqrt{2}$ to the linear element $2qp + 1$; it follows that the quadratic simple module $A/A(p^2 - q^2)$ is Ore.

See Dixmier [2] for a description of the automorphisms of A. We also recall his concepts of strictly nilpotent and strictly semisimple elements of A. Rather than repeating the original definition, we quote his characterization in terms of automorphisms ([1], 9.1-9.3): an element of A is strictly nilpotent [strictly semisimple] iff it is the image, under an automorphism of A, of a polynomial in q only [of an element $\lambda qp + \mu$ with $\lambda \neq 0$, $\mu \in C$].

With some care to detail, one can deduce from Theorem 7 (cf. [8]):

Corollary 9. Let S be a simple A-module of degree ≥ 2. Assume that $[S] \cap A$ contains a member which is (i) of total degree ≤ 2, (ii) strictly nilpotent, or (iii) strictly semisimple. Then S is Ore.

6. SUPPLEMENTS

There is a different, and much cheaper, way of producing Ore localizations in A:

It is routine to check that, in arbitrary rings, the iteration of two Ore localizations is again an Ore localization. Now B is an Ore localization of A. But B is a principal ideal domain and, therefore, every localization of B is Ore ([7], XI.6.1). We deduce

Lemma 10. Let γ be a class of simple A-modules which contains all B-torsion ones. Then γ is Ore.

As the Ore condition is preserved under automorphisms, and the image of the B-torsion simple module $A/A(q - \rho)$ under the automorphism σ is $A/A(\sigma q - \rho)$, we obtain

Corollary 11. If γ contains all $A/A(\sigma q - \rho)$, $\rho \in C$, for some automorphism σ of A, then γ is Ore.

Again with a little care, and utilizing an automorphism $q \to q$, $p \to p + f(q)$ for a suitable polynomial f, one proves (cf. [8]):

Corollary 12. If γ contains all but countably many isomorphism types of simple A-modules, then γ is Ore.

REFERENCES

1. R.E. Block, The irreducible representations of the Lie algebra sl(2) and of the Weyl algebra, Advances in Math. **39** (1981), 69-110.
2. J. Dixmier, Sur les algèbres de Weyl, Bull. Soc. Math. France **96** (1968), 209-242.
3. J.S. Golan, *Localization of Noncommutative Rings*, Marcel Dekker Inc. (1975).
4. K.R. Goodearl, Linked injectives and Ore localizations, J. London Math. Soc. **37** (1988), 404-420.
5. J.C. McConnell and J.C. Robson, Homomorphisms and extensions of modules over certain differential polynomial rings, J. Algebra **26** (1973), 319-342.
6. J.C. McConnell and J.C. Robson, *Noncommutative Noetherian rings*, John Wiley and Sons (1987).
7. B. Stenström, *Rings of quotients*, Springer (1975).
8. Y.L. Zhang, Ore localizations and irreducible representations of the first Weyl algebra, Ph.D. thesis, McMaster University (1990).

McMaster University
Hamilton, Ontario L8S 4K1
CANADA

and

Academia Sinica
Beijing 100080
PEOPLE's REPUBLIC OF CHINA
and McMaster University

RING-THEORETICAL ASPECTS OF THE BERNSTEIN-BEILINSON THEOREM

Timothy J. Hodges*

INTRODUCTION

The recent proof of the Kazhdan-Lusztig conjectures involved a number of deep results from some diverse areas of mathematics: intersection cohomology, D-modules, holonomic systems with regular singularities, perverse sheaves and representation theory of Lie algebras. One of the key results that ties together these ideas is the Bernstein-Beilinson theorem proved in 1980. This result established an equivalence of categories between the category of modules over certain primitive factors of the enveloping algebra of a semi-simple Lie algebra, and the category of quasi-coherent modules over a related twisted sheaf of differential operators on the associated flag variety. This result allows one to apply the powerful machinery of algebraic geometry, intersection cohomology, D-modules, etc. to the study of the representations of the Lie algebra. The proof of this theory by Bernstein and Beilinson (and simultaneously, in a slightly different form by Brylinski and Kashiwara) completed the proof of the Kazhdan Lusztig conjecture.

In this article, we shall be concerned with the significance to ring theory of the Bernstein-Beilinson theorem. We shall take it apart and look at it from a ring theorist's perspective and then ask what there is to learn from it. In particular, I shall try to suggest that this result is not an isolated phenomenon but a behavior that happens in a number of different situations. The key ingredient is torsion-theoretic localization and the message that comes across is that this kind of localization is a fundamental and very natural tool in the study of noncommutative algebras. While classical elementwise localization is a strong enough tool to do algebraic geometry, the additional complications in the noncommutative setting make such Ore localizations rarer and put more emphasis on the more general process.

In the third section we discuss the possibility of a "quantum analog" of the Bernstein-Beilinson theorem for the q-analogs of the enveloping algebra of a semi-simple Lie algebra. As experimental evidence we sketch a proof of the result for $U_q(\mathfrak{sl}(2, \mathbf{C}))$.

1. THE BERNSTEIN-BEILINSON THEOREM

To avoid a deluge of definitions and technicalities, I shall say little more of the subjects mentioned in introduction, not even stating the Kazhdan-Lusztig conjecture. Moreover I will not even define in any detail the concepts involved in the Bernstein-Beilinson theorem. Rather I shall try to convey a rough idea of what is going on. The interested reader may refer to more comprehensive accounts for the precise definitions. In particular we would like to mention the notes of Dragan Miličić [M] which provide a beautifully readable account of this subject. The original paper gives a very concise statement of the theorem and its consequences.

------

* The author was partially supported by a joint grant from the National Science Foundation and the National Security Agency (MDA 904-89-H2046).

In order to state the theorem, we will need a certain amount of fairly standard notation. Detailed definitions may be found in Dixmier's book [D] or in Miličić's notes. Let $\mathfrak{g}$ be a finite dimensional semi-simple Lie algebra ($\mathfrak{sl}(n, \mathbf{C})$ in the canonical example) and let $\mathfrak{h}$ be a Cartan subalgebra (for $\mathfrak{sl}(n, \mathbf{C})$, $\mathfrak{h}$ is just the subalgebra of diagonal matrices). The most important $\mathfrak{g}$-modules are the Verma modules $M(\lambda)$, for $\lambda \in \mathfrak{h}^*$. These modules are $\mathfrak{h}$-diagonalizable and have finite length and unique simple quotients $L(\lambda)$. The primitive ideals of the enveloping algebra $U(\mathfrak{g})$ are precisely the ideals Ann $L(\lambda)$. On the other hand, the minimal primitive ideals of $U(\mathfrak{g})$ may also be described as the ideals Ann $M(\lambda)$ where λ varies over $\mathfrak{h}^*$. There are notions of regularity and dominance for elements of $\mathfrak{h}^*$ which we will not discuss here. Suffice it to say that the interesting cases are covered by the assumption that λ is dominant and regular.

So much for the terminology from enveloping algebras. Now let us turn our attention to differential operators and "$\mathcal{D}$-modules". For a smooth affine algebraic variety X, the ring of differential operators $\mathcal{D}(X)$ is the algebra generated by the regular functions $\mathcal{O}(X)$ and the derivations Der (X). The sheaf of differential operators on a general quasi-projective variety is the sheaf $\mathcal{D}$ whose local sections are the algebras $\mathcal{D}(U)$. A twisted sheaf of differential operators is a sheaf $\mathcal{D}'$ that is locally isomorphic (in a controlled fashion) to $\mathcal{D}$. Now let X be the flag variety G/B associated to the semi-simple Lie algebra. (If $\mathfrak{g} = \mathfrak{sl}(n, \mathbf{C})$, then $G = SL(2, \mathbf{C})$, B is the Borel subgroup of upper triangular matrices and G/B is the usual variety of full flags as defined in [Hu]). It turns out that each $\lambda \in \mathfrak{h}^*$ defines a twisted sheaf $\mathcal{D}_\lambda$ of differential operators on G/B. Let us denote by D_λ the global sections $\Gamma(G/B, \mathcal{D}_\lambda)$.

We are now in a position to state the result.

Theorem 1. (Bernstein-Beilinson [BB])

(i) $D_\lambda = \Gamma(G/B, \mathcal{D}_\lambda) \cong U(\mathfrak{g})/$ Ann $M(\lambda)$.

(ii) If λ is dominant and regular, then the global section functor Γ defines an equivalence of categories between the category of quasi-coherent left $\mathcal{D}_\lambda$-modules and the category of left D_λ-modules.

At first glance it seems unlikely that this result could have any interpretation in purely ring-theoretical terms. However, this is the case, at least in part. For the following fact follows fairly directly from the theorem.

(*) Let $U_1, \cdots, U_t$ be an affine open cover for G/B. Then the rings $\mathcal{D}_\lambda(U_i)$ are perfect left localizations of D_λ. Furthermore $\bigoplus_{i=1}^{t} \mathcal{D}_\lambda(U_i)$ is a faithfully flat right D_λ-module.

Moreover, it was shown in [HS1] that the theorem follows formally from this information. Thus the existence of such an equivalence of categories is in some sense equivalent to assertion (*). Now flag varieties have the very pleasant property that they can be covered by copies of affine space. Thus there exists a cover $U_1, \cdots, U_t$ of G/B with $U_i \cong A^n(\mathbf{C})$ for all i (here $n = \dim G/B$). Moreover, it follows easily from the definition of twisted sheaves of differential operators that in this situation (i.e., when U_i is isomorphic to affine space), $\mathcal{D}(U_i)$ must be the usual ring of differential operators on U_i. That is, $\mathcal{D}(U_i)$ is isomorphic to the $n - th$ Weyl algebra $A_n(\mathbf{C})$. Hence we have the following corollary.

Corollary 2. Suppose that λ is dominant and regular. Then there exist copies of the Weyl algebra, $B_1, \cdots, B_t$ such that:

(i) Each B_i is a perfect left localization of D_λ.

(ii) $\displaystyle\bigoplus_{i=1}^{t} B_i$ is faithfully flat as a right D_λ-module.

Thus D_λ is, in a sense, locally isomorphic to the Weyl algebra. As such it shares many properties with the Weyl algebra. It must have the same Gelfand-Kirillov and Krull dimensions and the global dimension may be calculated using a little additional information. It also follows that D_λ must be a maximal order. (Of course, this is not always the easiest way to prove these results for D_λ).

So $A_n(\mathbf{C})$ is, at least in a torsion-theoretic sense, a localization of D_λ. Could it be a classical Ore localization? Obviously not, since the only units of $A_n(\mathbf{C})$ are the scalars. Exactly how to describe the associated torsion theory intrinsically is a tricky problem. However, in the next section we give a very simple description in the case $\mathfrak{g} = \mathfrak{sl}(2, \mathbf{C})$.

2. THE CASE $\mathfrak{g} = \mathfrak{sl}(2, \mathbf{C})$

In this section we consider in greater detail the case when $\mathfrak{g} = \mathfrak{sl}(2, \mathbf{C})$. In particular we indicate how to prove the above corollary by purely algebraic means. In this situation, the flag variety G/B is just the projective line $\mathbf{P}^1$. Let U_1, U_2 be the standard open affine cover for $\mathbf{P}^1$, and let q be a local parameter for U_1. Then the sheaf of differential operators is given locally by:

$$\Gamma(U_1, \mathcal{D}) = \mathbf{C}[p, q]$$

where $pq - qp = 1$

$$\Gamma(U_2, \mathcal{D}) = \mathbf{C}[q^2 p, q^{-1}] \ .$$

Now $\mathfrak{h}$ is one dimensional in this situation, so we may identify $\mathfrak{h}^*$ with $\mathbf{C}$. The twisted sheaves of differential operators may then be constructed by twisting slightly the position of the local sections on U_2. If we set

$$\Gamma(U_1, \mathcal{D}_\lambda) = C[p, q]$$

$$\Gamma(U_2, \mathcal{D}_\lambda) = \mathbf{C}[q(qp + \lambda), q^{-1}]$$

then this information is enough to define completely the sheaf $\mathcal{D}_\lambda$. The global sections are just $\Gamma(U_1, \mathcal{D}_\lambda) \cap \Gamma(U_2, \mathcal{D}_\lambda)$ and the natural map from $U(\mathfrak{g})$ to the global section is given by:

$$E \to -p \qquad F \to q(qp + \lambda)H \to -(2qp + \lambda) \ .$$

Denote the images of E, F and H by e, f and h, respectively, and denote the image of $U(\mathfrak{g})$ by $\bar{U}$. Let us look first at the embedding of $\bar{U}$ into $\Gamma(U_1, \mathcal{D}_\lambda)$.

Proposition 3.

(i) $\Gamma(U_1, \mathcal{D}_\lambda)$ is contained between $\bar{U}$ and its quotient division ring.

(ii) If λ is not a positive integer, then $\Gamma(U_1, \mathcal{D}_\lambda)$ is a flat right $\bar{U}$-module and hence is a perfect left localization of $\bar{U}$. The class of torsion modules is generated by the simple Verma module $M(\lambda - 1)$ of highest weight $\lambda - 2$.

Proof.

(i) is obvious.

(ii) Let $C_k = \bar{U} + q\bar{U} + \cdots + q^k\bar{U}$. It is clear that $\Gamma(U_1, \mathcal{D}_\lambda) = \cup_{k=1}^{\infty} C_k$. To prove flatness, it suffices to show that C_k is a projective right $\bar{U}$-module. This may be done by finding an explicit dual basis. For instance, for $k = 1$, note that p and $qp + \lambda - 1$ lie in C_1^*. Since $1.(qp + \lambda - 1) - q.p = \lambda - 1 \neq 0$, it follows that C_1 is projective.

It is a standard consequence of (i) and the flatness of $\Gamma(U_1, \mathcal{D}_\lambda)$ that $\Gamma(U_1, \mathcal{D}_\lambda)$ is a perfect left localization of $\bar{U}$ [S]. Since $\Gamma(U_1, \mathcal{D}_\lambda)$ is generated over $\bar{U}$ by q, the associated class of torsion modules must be generated by the module $M = (\bar{U} + \bar{U}q)/\bar{U}$ and M is easily seen to be the simple Verma module of highest weight $\lambda - 2$.

No further work is needed in order to understand the embedding of $\bar{U}$ into $\Gamma(U_2, \mathcal{D}_\lambda)$. For there is a standard isomorphism of $\bar{U}$ given by sending

$$e \to f \ , \quad f \to e \quad \text{and} \quad h \to -h$$

and this isomorphism extends to an isomorphism of the quotient division ring that maps $\Gamma(U_1, \mathcal{D}_\lambda)$ isomorphically onto $\Gamma(U_2, \mathcal{D}_\lambda)$. Furthermore the associated torsion class is now generated by the "opposite" Verma module; that is, the Verma module corresponding to the Borel subalgebra $\mathbf{C}F + \mathbf{C}H$ rather than $\mathbf{C}E + \mathbf{C}H$. Both these modules are h-diagnoalizable but with different weights. Therefore, they are certainly not isomorphic. Given this, the two torsion classes can have no modules in common. Thus $\Gamma(U_1, \mathcal{D}_\lambda) \cap \Gamma(U_2, \mathcal{D}_\lambda) = \bar{U}$ and $\Gamma(U_1, \mathcal{D}_\lambda) \oplus \Gamma(U_2, \mathcal{D}_\lambda)$ is a faithfully flat right $\bar{U}$-module. Thus we have proved Corollary 2 above in this case and in combination with the main result of [HS1] provides a ring theoretic proof of the Bernstein-Beilinson theorem for $\mathfrak{sl}(2, \mathbf{C})$. We state this result explicitly for comparison with results in the next section. The only infinite-dimensional primitive factor of $U(\mathfrak{sl}(2, \mathbf{C}))$ not covered by the above argument is the unique primitive factor with infinite global dimension.

Corollary 4. Let $\bar{U}$ be a regular infinite dimensional primitive factor of $U(\mathfrak{g})$. Then there exist copies B_1 and B_2 of the Weyl algebra such that B_1 and B_2 are perfect left localization of $\bar{U}$ and $B_1 \oplus B_2$ is a faithfully flat right $\bar{U}$-module.

It should be noted at this stage that this ring theoretic approach does not seem to extend easily to the general situation because of the difficulty in identifying the associated torsion class.

Examples of Similar Behavior

As I said in the introduction, this is not an isolated example of this kind of behavior and in the next section, I will discuss the situation for the q-analog $U_q(\mathfrak{sl}(2,\mathbf{C}))$ of the enveloping algebra $U(\mathfrak{sl}(2,\mathbf{C}))$. However, before moving on to look at these algebras, let us look at another class of algebras more familiar to algebraists. Let $A = A_1(\mathbf{C})$ be the usual Weyl algebra, let G be a cyclic group of order n acting in the standard diagonal fashion; that is, G is generated by the automorphism sending $p \to \omega p$ and $q \to \omega^{-1} q$ where ω is a primitive $n-th$ root of unity. An exactly analogous result is true for the fixed rings A^G.

Theorem 5 [H]. There exist n copies of the Weyl algebra, $B_1, \cdots, B_n$ such that:

(i) Each B_i is a perfect left localization of A^G.

(ii) $\displaystyle\bigoplus_{i=1}^{n} B_i$ is faithfully flat as a right A^G-module.

This result is much less surprising when one observes that in the case $G = \mathbf{Z}_2$, the fixed ring A^G is a particular primitive factor of $U(\mathfrak{sl}(2,\mathbf{C}))$. In fact, one can prove a more general result which includes both this result and Corollary 4. For the fixed rings A^G form part of a continuous family of deformations which include all the primitive factors of $U(\mathfrak{sl}(2,\mathbf{C}))$ and for which an analogous theorem holds [H]. There are interesting connections between this result and the desingularization of the associated graded ring $\mathbf{C}[X^n, XY, Y^n]$, and this leads me to conjecture that a similar result should be true for the other fixed rings of $A_1(\mathbf{C})$.

There is also a version of the Bernstein-Beilinson theorem for Lie superalgebras and flag supermanifolds (see [PS] and [P]). Presumably there is an analogous algebraic interpretations of these results. A version of the classical result also exists for fields of finite characteristic [Ha] through the results here are much weaker.

3. A BERNSTEIN-BEILINSON THEOREM FOR $U_q(\mathfrak{sl}(2,\mathbf{C}))$

A large number of results concerning algebraic groups and their associated Lie algebras have analogs in the theory of quantum groups. Given the importance of the Bernstein-Beilinson theorem in the classical case, it is natural to ask if an analog of this result holds for quantum groups. In the case of $U_q(\mathfrak{sl}(2,\mathbf{C}))$, the answer is "yes" and what follows is an outline of this result. For any unexplained notation the reader is referred to the survey article of S.P. Smith [S2].

Recall the definition of $U_q(\mathfrak{sl}(2,\mathbf{C}))$. It is the $\mathbf{C}$-algebra generated by elements E, F and $K^{\pm 1}$ subject to the relations:

$$KE = q^2 EK \;,\; KF = q^{-2} FK \text{ and } EF - FE = (K^2 - K^{-2})/(q^2 - q^{-2})\;.$$

Here q is an element of **C** which, in the following, will be assumed not to be a root of unity.

U_q has a Hopf algebra structure where the comultiplication is given by:

$$\Delta(E) = E \otimes K^{-1} + K \otimes E \ , \ \Delta(F) = F \otimes K^{-1} + K \otimes F \ , \ \Delta(K) = K \otimes K \ .$$

In the classical situation $U(\mathfrak{g})$ acts on the rings of local sections $\Gamma(U, \mathcal{O}(G/B))$ in such a way that $\Gamma(U, \mathcal{O}(G, B))$ becomes a left Hopf module algebra over $U(\mathfrak{g})$. Thus it is natural to ask whether the polynomial ring **C**$[T]$ can be made into a module algebra over U_q. One is immediately drawn into considering skew derivations of the following kind. Let R be a ring and let σ be an automorphism of R. We shall say that an endomorphism δ of R is a σ-derivation if $\sigma(ab) = \delta(a)\sigma^{-1}(b) + \sigma(a)\delta(b)$. (Note that this is not the usual definition of a σ-derivation.) If B is any module algebra over U_q, then K must act as an automorphism and E and F as K-derivations. Let σ be the automorphism of **C**$[T]$ given by $T \to q^2T$ and let δ be the σ-derivation given by $\delta(T) = 1$. Let A_q be the subalgebra of End $_\mathbf{C}$**C**$[T]$ generated by T, δ and σ. Then A_q seems to be the natural q-analog of the Weyl algebra and there is a map from U_q to A_q given by

$$\psi(E) = \delta \ , \ \psi(F) = -q^{-4}T^2\delta \ , \ \psi(K) = \sigma$$

(see [MS]). Thus A_q plays the role of $\Gamma(U_1, D)$. The natural candidate for the other ring of local sections is the image of A_q under the automorphism of its quotient division ring given by sending T to T^{-1}, namely $A'_q = \mathbf{C}[T^{-1}, T^2\delta, \sigma^{\pm 1}]$. These are exactly the algebras we need to prove a version of Corollary 4 for $\psi(U_q)$.

Theorem 6. The algebras A_q and A'_q are perfect left localizations of $\psi(U_q)$ and $A_q \oplus A'_q$ is a faithfully flat right $\psi(U_q)$-module.

The proof proceeds along the same lines as the proof of Proposition 3 and Corollary 4. One shows explicitly that A_q is a union of finitely generated projective submodules. The torsion class may then be identified as being generated by a certain analog of a Verma module and faithfulness follows easily.

Of course, this only gives an analog for the untwisted part of Corollary 4. In order to prove the more general result, we need to take

$$A'_q = \mathbf{C}[T^{-1}, T((1 - \lambda^4)(1 - q^4) - q^{-4}\lambda^4 T\delta), \sigma^{\pm 1}] \ .$$

However, the computations now become extremely laborious and one is naturally led to asking whether there is some way of using the kind of geometrical tools used to prove the Bernstein-Beilinson theorem. It turns out that there is, and the constructions involved show that A_q and A'_q arise very naturally. What follows is a rough outline of this approach. In order to fully understand what is going on, the reader should be reasonably familiar with the construction of twisted sheaves of differential operators as described, for instance, in [M] and with the basic constructions from the theory of quantum groups as outlined in [S2]. A complete account of this result will appear elsewhere.

Define $\mathcal{O}_q(\mathbf{C}^2)$ (the "coordinate ring of the quantum plane") to be the algebra generated by T_1 and T_2 subject to $T_1 T_2 = q^2 T_2 T_1$. Denote by X the space Proj $\mathcal{O}_q(\mathbf{C}^2)$ of homogeneous prime ideals (with respect to the obvious grading) with the usual Zariski topology. Since the only nonredundant homogeneous prime ideals of $\mathcal{O}_q(\mathbf{C}^2)$ are those generated by the T_i, X is a finite space with only three nontrivial open sets, U_1, U_2 and $U_1 \cap U_2$ where $U_i = X - \{(T_i)\}$. Although X is small, it is big enough to do some significant (noncommutative) geometry. First define the sheaf $\mathcal{T}$ of homogeneous functions of X by

$$\mathcal{T}(U_i) = \mathbf{C}[T_1, T_2, T_i^{-1}] .$$

Then $\mathcal{T}$ is a sheaf of graded algebras (graded by total degree in T_1 and T_2). Define $\mathcal{O}_q$ to be $\mathcal{T}_0$, the sheaf of degree zero terms. This sheaf plays the role of "regular functions on the quantum projective line", and it is very closely related to the usual sheaf of regular functions on the projective line. The invertible sheaves corresponding to line bundles are then naturally defined to be the $\mathcal{O}_q$ bimodules $\mathcal{O}_q(n) = \mathcal{T}_n$, the sheaves of elements of degree n.

This deals with the underlying geometry. Next one has to deal with the analog of differential operators. As we said above, the Hopf algebra structure of $U_q(\mathfrak{sl}(2, \mathbf{C}))$ defines exactly what this analog should be. In fact, at this stage, one has merely to reformulate the construction of twisted sheaves of differential operators into the language of Hopf algebras and the construction goes through *mutatis mutandum*. There is a natural action of $U_q(\mathfrak{sl}(2, \mathbf{C}))$ on $\mathcal{O}_q(\mathbf{C}^2)$ which makes $\mathcal{O}_q(\mathbf{C}^2)$ into a Hopf module algebra, and this action extends to $\mathcal{T}$ making $\mathcal{T}$ into a sheaf of Hopf module algebras. Moreover, $\mathcal{O}_q$ is a sheaf of submodule algebras. Therefore, we may form the smashed product sheaves,

$$\mathcal{T} \# U_q(\mathfrak{sl}(2, \mathbf{C})) \quad \text{and} \quad \mathcal{U}^0 = \mathcal{O}_q \# U_q(\mathfrak{sl}(2, \mathbf{C})) .$$

Henceforth we shall denote $U_q(\mathfrak{sl}(2, \mathbf{C}))$ by U_q. There is a natural map from $\mathcal{T} \# U_q$ to $\mathcal{E}nd_{\mathbf{C}} \mathcal{T}$. Denote the image of restriction of this map to $\mathcal{U}^0$ by $\mathcal{D}_K$. The sheaf $\mathcal{D}_K$ is a kind of generic sheaf of twisted operators. The kernel of this restriction turns out to be $\mathcal{U}^0 \otimes_{\mathcal{O}_q} \mathcal{O}_q(-2)$, yielding the following diagram.

$$
\begin{array}{ccc}
\mathcal{T} \# \mathcal{O}_q & \longrightarrow & \mathcal{E}nd_{\mathbf{C}} \mathcal{T} \\
\uparrow & & \uparrow \\
\end{array}
$$

$$0 \longrightarrow \mathcal{U}^0 \otimes_{\mathcal{O}_q} \mathcal{O}_q(-2) \longrightarrow \mathcal{U}^0 \longrightarrow \mathcal{D}_K \longrightarrow 0$$

Theorem 7.

(i) $\Gamma(X, \mathcal{D}_K) \cong U_q \otimes_{Z_q} \mathbf{C}[K^{\pm 2}]$

(ii) $H^i(X, \mathcal{D}_K) = 0$ for $i > 0$.

Note: H^i refers to the derived functors of the usual global section functor Γ. The tensor product in part (i) is over the center Z_q of U_q via the q-analog $\chi : Z_p \to \mathbf{C}[K^{\pm 1}]$ of the Harish-Chandra homomorphism [R].

Proof. The main components of the proof are the facts that

$$H^i(X, \mathcal{U}^0 \otimes \mathcal{O}_q(n)) \cong U_q \otimes_{\mathbf{C}} H^i(X, \mathcal{O}_q(n))$$

and that

$$H^i(X, \mathcal{O}_q(n)) \cong H^i(\mathbf{P}^1(\mathbf{C}), \mathcal{O}(n)) .$$

The theorem then follows from standard homological algebra and the fact that $U_q \otimes_{Z_q} \mathbf{C}[K^{\pm 2}]$ is a maximal order in the sense of [MR; 5.1.1].

Now let $\lambda : \mathbf{C}[K^{\pm 2}] \to \mathbf{C}$ be an algebra map and let $I_\lambda = Ker \, \lambda \cap \mathbf{C}[K^{\pm 2}]$. Define

$$\mathcal{D}_\lambda = \mathcal{D}_K / I_\lambda \mathcal{D}_K .$$

This is the required q-analog of a twisted sheaf of differential operators. clearly we have an exact sequence,

$$0 \to \mathcal{D}_K \to \mathcal{D}_K \to \mathcal{D}_\lambda \to 0 .$$

Theorem 8.

(i) $\Gamma(X, \mathcal{D}_\lambda) \cong U_q/(\mathcal{X}^{-1}(I_\lambda))$.

(ii) If λ is dominant and regular, then the global section functor Γ defines an equivalent of categories from the category of quasi-coherent left $\mathcal{D}_\lambda$-modules to the category of left modules over $U_q/(\chi^{-1}(I_\lambda))$.

Proof. The proof of part (i) is standard homological algebra. The proof of part (ii) follows closely the proof in the classical case [BB].

Again we may deduce the standard ring-theoretical corollary.

Corollary 9. Let $\bar{U}_q$ be a regular infinite dimensional primitive factor of U_q. Then there exist algebras B_1 and B_2 such that

(i) $B_i \cong A_q$

(ii) $B_1 \oplus B_2$ is faithfully flat as a $\bar{U}_q$-module

(iii) B_i is a perfect left localization of $\bar{U}_q$.

There are many obstacles to extending this result to the general situation, most notably the lack of a developed theory of the geometry of the quantum flag variety. However, despite these complications, I conjecture that some kind of close analog of the Bernstein-Beilinson theorem will hold in general.

REFERENCES

[BB] I.N. Bernstein and A. Beilinson, "Localisation de g-modules", C.R. Acad. Sci. Paris Sér. I 292 (1981) 15-18.

[BK] J.L. Brylinski and M. Kashiwara, "Kazhdan-Lusztig conjecture and holonomic systems", Invent. Math. 64 (1981) 387-410.

[Ha] B. Haastert, "Uber differentialoperatoren und $\mathbf{D}$-moduln in positiver charakteristik", Manuscripta math. 58 (1987) 385-415.

[H] T.J. Hodges, "Noncommutative deformations of type A Kleinian singularities", preprint, University of Cincinnati, 1990.

[HS1] T.J. Hodges and S.P. Smith, "Rings of differential operators and the Bernstein-Beilinson equivalence of categories", Proc. Amer. Math. Soc. 93 (1985) 379-386.

[HS2] T.J. Hodges and S.P. Smith, "The global dimension of certain primitive factors of the enveloping algebra of a semi-simple Lie algebra", J. London Math. Soc. (2), 82 (1985) 411-418.

[Hu] J.E. Humphreys, *Linear algebraic groups,* Springer-Verlag, New York, 1981.

[MR] J.C. McConnell and J.C. Robson, *Noncommutative Noetherian rings,* Wiley, Chichester, 1987.

[M] D. Miličić, "Localization and representation theory of reductive Lie groups", Lecture Notes, University of Utah, 1987, to appear.

[MS] S. Montgomery and S.P. Smith, "Skew derivations and $U_q(sl(2, \mathbf{C}))$", Israel J. Math., to appear.

[PS] I. Penkov and I. Skornyakov, "Cohomologie des D-modules tordus typiques sur les supervariétés de drapeaux", C.R. Acad. Sci. Paris, Ser. I, 299 (1984) 1005-1008.

[P] I. Penkov, "Localisation des représentations typiques d'une superalgèbre de Lie complexe classique", C.R. Acad. Sci. Parais, Ser. I, 304 (1987) 163-166.

[R] M. Rosso, "Analogues de la forme de Killing et du théorème de Harish-Chandra pour les groupes quantiques", preprint, Palaiseau, 1989.

[S1] S.P. Smith, "A class of algebras similar to the enveloping algebra of $U(sl(2, \mathbf{C}))$", Pac. J. Math., to appear.

[S2] S.P. Smith, "Quantum groups: An introduction and survey for ring theorists", preprint, University of Washington, 1990.

[S] B. Stenstrom, *Rings of quotients,* Springer-Verlag, Berlin, 1975.

University of Cincinnati
Cincinnati, Ohio 45221-0025

LIST OF PARTICIPANTS

AFTABIZADEH, REZA , Ohio University, Athens, OH

AKIRA, ABE , Indiana University, Bloomington, IN

ALFARO, RICARDO , University of Michigan-Flint, Flint, MI

AL-HUZALI, ABDULLAH , Ohio University, Athens, OH

AZUMAYA, GORO , Indiana University, Bloomington, IN

BEHR, ERIC , Illinois State University-Normal, Normal, IL

BIRKENMEIER, GARY , University of Southwestern Louisiana, Lafayette, LA

BURKHOLDER, DOUGLAS , Wichita State University, Wichita, KS

CAMILLO, VICTOR , University of Iowa, Iowa City, IO

CORTZEN, BARBARA , DePaul University, Chicago, IL

DAUNS, JOHN , Tulane University, New Orleans, LA

ELDRIDGE, KLAUS , Ohio University, Athens, OH

FERRAR, JOE , Ohio State University, Columbus, OH

FORMANEK, ED , Pennsylvania State University, University Park, PA

FULLER, KENT , University of Iowa, Iowa City, IO

HODGES, TIM , University of Cincinnati, Cincinnati, OH

JANSEN, WILLEM , University of Toledo, Toledo, OH

JAIN, S.K., Ohio University, Athens, OH

KOSLER, KARL , University of Wisconsin-Waukesha, Waukesha, WI

KOKER, JOHN , University of Wisconsin-Milwaukee, Milwaukee, WI

KHURI, SOUMAYA, East Carolina University, Greenville, NC

LI, M.S., Oberlin College, Oberlin, OH

LÓPEZ-PERMOUTH, SERGIO R., Ohio University, Athens, OH

LOUSTAUNAU, PHILIPPE , George Mason University, Fairfax, VA

McCONNEL, JOHN C., University of Leeds, Leeds, ENGLAND

MALM, DENNIS , Northwest Missouri State University, Maryville, MO

MING-SUN, LI , Oberlin College, Oberlin, OH

MÜLLER, BRUNO , McMaster University, Hamilton, Ontario, CANADA

NIJASURE, SHUBHANGI , University of Wisconsin-Milwaukee, Milwaukee, WI

OKNINSKI, JAN , North Carolina State University, Raleigh, NC

OLSON, DWIGHT , John Carroll University, University Heights, OH

OSTERBURG, JIM , University of Cinncinati, Cincinnati, OH

REDMAN, DAVID B. (Jr.), University of Wisconsin-Milwaukee, Milwaukee, WI

RIM, SEOG HOON , University of Wisconsin-Milwaukee, Milwaukee, WI

RIZVI , S. TARIQ , Ohio State University-Lima, Lima, OH

SALEH, MOHAMMAD , Ohio University, Athens, OH

SHAPIRO, DANIEL , Ohio State University, Columbus, OH

SHAPIRO, JAY , George Mason University, Fairfax, VA

SMITH, P.F., University of Glasgow, Glasgow, SCOTLAND

SNYDER, LARRY , Ohio University, Athens, OH

TEPLY, MARK , University of Wisconsin-Milwaukee, Milwaukee, WI

VANCKO, ROBERT , Ohio University, Athens, OH

WEN, SHIH , Ohio University, Athens, OH

WEHLEN, JOE , Computer Sciences Corporation, Moorestown, NJ

WRIGHT, MARY , University of Southern Illinois, Carbondale, IL

YOUSIF, MOHAMED F., Ohio State University-Lima, Lima, OH

ZHU, RUIYING , Ohio University, Athens, OH

ecture Notes aim to report new developments - quickly, informally nd at a high level. The following describes criteria and procedures hich apply to proceedings volumes. The editors of a volume are trongly advised to inform contributors about these points at an arly stage.

1. One (or more) expert participant(s) of the meeting should act as the responsible editor(s) of the proceedings. They select the papers which are suitable (cf. §§ 2, 3) for inclusion in the proceedings, and have them individually refereed (as for a journal). It should not be assumed that the published proceedings must reflect conference events faithfully and in their entirety. Contributions to the meeting which are not included in the proceedings can be listed by title. The series editors will normally not interfere with the editing of a particular proceedings volume - except in fairly obvious cases, or on technical matters, such as described in §§ 2, 3. The names of the responsible editors appear on the title page of the volume.

2. The proceedings should be reasonably homogeneous (concerned with a limited area). For instance, the proceedings of a congress on "Analysis" or "Mathematics in Wonderland" would normally not be sufficiently homogeneous.

   One or two longer survey articles on recent developments in the field are often very useful additions to such proceedings - even if they do not correspond to actual lectures at the congress. An extensive introduction on the subject of the congress would be desirable.

3. The contributions should be of a high mathematical standard and of current interest. Research articles should present new material and not duplicate other papers already published or due to be published. They should contain sufficient information and motivation and they should present proofs, or at least outlines of such, in sufficient detail to enable an expert to complete them. Thus resumes and mere announcements of papers appearing elsewhere cannot be included, although more detailed versions of a contribution may well be published in other places later.

   Contributions in numerical mathematics may be acceptable without formal theorems resp. proofs if they present new algorithms solving problems (previously unsolved or less well solved) or develop innovative qualitative methods, not yet amenable to a more formal treatment. .

   Surveys, if included, should cover a sufficiently broad topic, and should in general not simply review the author's own recent research. In the case of such surveys, exceptionally, proofs of results may not be necessary.

4. "Mathematical Reviews" and "Zentralblatt für Mathematik" recommend that papers in proceedings volumes carry an explicit statement that they are in final form and that no similar paper has been or is being submitted elsewhere, if these papers are to be considered for a review. Normally, papers that satisfy the criteria of the Lecture Notes in Mathematics series also satisfy

this requirement, but we strongly recommend that the contributing authors be asked to give this guarantee explicitly at the beginning or end of their paper. There will occasionally be cases where this does not apply but where, for special reasons the paper is still acceptable for LNM.

§5. Proceedings should appear soon after the meeeting. The publisher should, therefore, receive the complete manuscript (preferably in duplicate) within nine months of the date of the meeting at the latest.

§6. Plans or proposals for proceedings volumes should be sent to one of the editors of the series or to Springer-Verlag Heidelberg They should give sufficient information on the conference o symposium, and on the proposed proceedings. In particular, they should contain a list of the expected contributions with their prospective length. Abstracts or early versions (drafts) of some of the contributions are helpful.

§7. Lecture Notes are printed by photo-offset from camera-ready typed copy provided by the editors. For this purpose Springer-Verlag provides editors with technical instructions for the preparation of manuscripts and these should be distributed to all contributing authors. Springer-Verlag can also, on request supply stationery on which the prescribed typing area is outlined. Some homogeneity in the presentation of the contributions is desirable.

Careful preparation of manuscripts will help keep production time short and ensure a satisfactory appearance of the finished book. The actual production of a Lecture Notes volume normally takes 6 -8 weeks.

Manuscripts should be at least 100 pages long. The final version should include a table of contents.

§8. Editors receive a total of 50 free copies of their volume for distribution to the contributing authors, but no royalties. (Unfortunately, no reprints of individual contributions can be supplied.) They are entitled to purchase further copies of their book for their personal use at a discount of 33.3 %, other Springer mathematics books at a discount of 20 % directly from Springer-Verlag. Contributing authors may purchase the volume in which their article appears at a discount of 33.3 %.

Commitment to publish is made by letter of intent rather than by signing a formal contract. Springer-Verlag secures the copyright for each volume.

Addresses:

Professor A. Dold, Mathematisches Institut, Universität Heidelberg,
Im Neuenheimer Feld 288, 6900 Heidelberg, Federal Republic of Germany

Professor B. Eckmann, Mathematik, ETH-Zentrum
8092 Zürich, Switzerland

Prof. F. Takens, Mathematisch Instituut, Rijksuniversiteit Groningen,
Postbus 800, 9700 AV Groningen, The Netherlands

Springer-Verlag, Mathematics Editorial, Tiergartenstr. 17,
6900 Heidelberg, Federal Republic of Germany, Tel.: (06221) 487-410

Springer-Verlag, Mathematics Editorial, 175, Fifth Avenue,
New York, New York 10010, USA, Tel.: (212) 460-1596